Embedded Systems Interfacing for Engineers using the Freescale HCS08 Microcontroller II: Digital and Analog Hardware Interfacing

Synthesis Lectures on Digital Circuits and Systems

Editor
Mitchell A. Thornton, *Southern Methodist University*

Embedded Systems Interfacing for Engineers using the Freescale HCS08 Microcontroller II: Digital and Analog Hardware Interfacing
Douglas H. Summerville
2009

Designing Asynchronous Circuits using NULL Convention Logic (NCL)
Scott C. Smith, JiaDi
2009

Embedded Systems Interfacing for Engineers using the Freescale HCS08 Microcontroller I: Assembly Language Programming
Douglas H.Summerville
2009

Developing Embedded Software using DaVinci & OMAP Technology
B.I. (Raj) Pawate
2009

Mismatch and Noise in Modern IC Processes
Andrew Marshall
2009

Asynchronous Sequential Machine Design and Analysis: A Comprehensive Development of the Design and Analysis of Clock-Independent State Machines and Systems
Richard F. Tinder
2009

An Introduction to Logic Circuit Testing
Parag K. Lala
2008

Pragmatic Power
William J. Eccles
2008

Multiple Valued Logic: Concepts and Representations
D. Michael Miller, Mitchell A. Thornton
2007

Finite State Machine Datapath Design, Optimization, and Implementation
Justin Davis, Robert Reese
2007

Atmel AVR Microcontroller Primer: Programming and Interfacing
Steven F. Barrett, Daniel J. Pack
2007

Pragmatic Logic
William J. Eccles
2007

PSpice for Filters and Transmission Lines
Paul Tobin
2007

PSpice for Digital Signal Processing
Paul Tobin
2007

PSpice for Analog Communications Engineering
Paul Tobin
2007

PSpice for Digital Communications Engineering
Paul Tobin
2007

PSpice for Circuit Theory and Electronic Devices
Paul Tobin
2007

Pragmatic Circuits: DC and Time Domain
William J. Eccles
2006

Pragmatic Circuits: Frequency Domain
William J. Eccles
2006

Pragmatic Circuits: Signals and Filters
William J. Eccles
2006

High-Speed Digital System Design
Justin Davis
2006

Introduction to Logic Synthesis using Verilog HDL
Robert B.Reese, Mitchell A.Thornton
2006

Microcontrollers Fundamentals for Engineers and Scientists
Steven F. Barrett, Daniel J. Pack
2006

© Springer Nature Switzerland AG 2022

Reprint of original edition © Morgan & Claypool 2009

All rights reserved. No part of this publication may be reproduced, stored in a retrieval system, or transmitted in any form or by any means—electronic, mechanical, photocopy, recording, or any other except for brief quotations in printed reviews, without the prior permission of the publisher.

Embedded Systems Interfacing for Engineers using the Freescale HCS08 Microcontroller II:
Digital and Analog Interfacing
Douglas H. Summerville

ISBN: 978-3-031-79802-3   paperback
ISBN: 978-3-031-79803-0   ebook

DOI 10.1007/978-3-031-79803-0

A Publication in the Springer series
SYNTHESIS LECTURES ON DIGITAL CIRCUITS AND SYSTEMS

Lecture #22
Series Editor: Mitchell A. Thornton, *Southern Methodist University*

Series ISSN
Synthesis Lectures on Digital Circuits and Systems
Print 1932-3166   Electronic 1932-3174

Embedded Systems Interfacing for Engineers using the Freescale HCS08 Microcontroller II: Digital and Analog Hardware Interfacing

Douglas H. Summerville
State University of New York at Binghamton

SYNTHESIS LECTURES ON DIGITAL CIRCUITS AND SYSTEMS #22

ABSTRACT

The vast majority of computers in use today are encapsulated within other systems. In contrast to general-purpose computers that run an endless selection of software, these embedded computers are often programmed for a very specific, low-level and often mundane purpose. Low-end microcontrollers, costing as little as one dollar, are often employed by engineers in designs that utilize only a small fraction of the processing capability of the device because it is either more cost-effective than selecting an application-specific part or because programmability offers custom functionality not otherwise available. *Embedded Systems Interfacing for Engineers using the Freescale HCS08 Microcontroller* is a two-part book intended to provide an introduction to hardware and software interfacing for engineers. Building from a comprehensive introduction of fundamental computing concepts, the book suitable for a first course in computer organization for electrical or computer engineering students with a minimal background in digital logic and programming. In addition, this book can be valuable as a reference for engineers new to the Freescale HCS08 family of microcontrollers. The HCS08 processor architecture used in the book is relatively simple to learn, powerful enough to apply towards a wide-range of interfacing tasks, and accommodates breadboard prototyping in a laboratory using freely available and low-cost tools.

In *Part II: Digital and Analog Hardware Interfacing*, hardware and software interfacing concepts are introduced. The emphasis of this work is on good hardware and software engineering design principles. Device drivers are developed illustrating the use of general-purpose and special-purpose digital I/O interfaces, analog interfaces, serial interfaces and real-time I/O processing. The hardware side of each interface is described and electrical specifications and related issues are considered. The first part of the book provides the programming skills necessary to implement the software in this part.

KEYWORDS

microcontrollers, embedded computers, computer engineering, digital systems, Freescale HCS08, device drivers, hardware/software interfacing

Contents

Acknowledgments

Most of all, I thank Erika, Nicholas, Daniel and Stephanie for their support, encouragement and for tolerating all the weekends and evenings that I spent working on this book instead of being with them. This book is dedicated to you. I also thank all the students who debugged the early versions of this book and whose curiosity, questions and comments helped to shape the presentation of this material.

Douglas H. Summerville
August 2009

CHAPTER 1

Introduction to the MC9S08QG4/8 Hardware

A computer is a device capable of processing data under control of a program. Input and output provide the means by which the computer can receive, exchange, or transmit such data. Without input/output (I/O), even the most powerful computer is all but useless. While general-purpose computers can be used to run a wide range of applications, embedded computers are primarily about I/O processing.

Embedded Systems Interfacing for Engineers using the Freescale HCS08 Microcontroller is a two-part book intended to provide an introduction to hardware and software interfacing concepts. In *Part I: Assembly Language Programming*, the programmer's model of the HSC08 family of processors is introduced, intended to prepare the engineer with the programming skills necessary to write device drivers and perform basic input/output processing in this part. The emphasis of Part II: *Digital and Analog Hardware Interfacing* is on hardware and software design concepts necessary to integrate hardware components into the embedded microcomputer system.

1.1 INPUT/OUTPUT BASICS

I/O refers collectively to the hardware and software methods used by a computer to interact with its environment. A computer's I/O unit is a collection of hardware interfaces used by the central processing unit (CPU) to interact with peripheral devices and other computers. An *interface* (or *port*) is a well-defined specification for communication between two devices, including the mechanical, electronic and data format standards. Generally, devices connected to these interfaces interact with people or other systems, or simply sense or control the environment.

1.1.1 PIN DIAGRAMS

The MC9S08QG family of microcontrollers from Freescale Semiconductor feature an easy to learn 8-bit architecture, up to 8 KiB of Flash ROM, up to 512B of RAM, and a rich set of peripherals including an analog to digital convertor, three serial communications interfaces (IIC, SCI and SPI), an analog comparator and pulse width modulator. In addition, these low-cost devices come in a variety of packages, including dual-inline packages that facilitate breadboard prototyping. In addition, low-cost development kits and free development software are available.

Pin diagrams of the MC9S08QG4 in an 8-pin DIP package and MC9S08QG8 in a 16-pin DIP package are shown in Figure 1.1. These devices come in a variety of other 8 and 16 pin packages. Pins 1 through 4 are identical on both packages; pins 5 through 8 on the MC9S08QG4 are equivalent

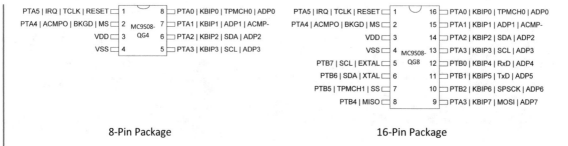

Figure 1.1: Device package pin assignments for the MC9S08QG4/8 microcontrollers.

to pins 13 through 16 on the MC9S08QG8. V_{DD} and V_{SS} (pins 3 and 4) are the power supply pins (power and ground, respectively). The manufacturer recommends that two separate capacitors be placed across the power pins. A 10-μF tantalum capacitor is recommended to provide bulk charge storage and a 0.1-μF ceramic bypass capacitor is recommended to suppress high-frequency noise; the latter should be placed as close to the power-supply pins as possible.

All pins other than power supply pins have multiple I/O functions assigned to them. Multiplexing pins in this way is a common technique used with microcontrollers to provide flexibility in assigning functions to pins and minimize unused pins. The system designer can select the functions needed in a particular system and configure the pins accordingly. When multiple I/O functions are enabled on a pin an assigned priority determines which interface uses that pin. For example, if the analog to digital convertor function is enabled on pin 13 (ADC3), that function is given priority over the GPIO interface on that pin (PTA3). This multiplexing approach maximizes the number of peripherals that can be included within a given package size. Since unused interfaces may not result in unused pins, it can also maximize pin utilization.

1.1.2   MEMORY-MAPPED I/O

I/O interfaces can be input only, output only or bidirectional (input and output). Generally, an I/O interface includes of a set of registers through which the CPU can read or write data. These I/O registers can be classified into one of three types: data, control and status. A *data register* is used for exchanging data with the interface, a *control register* is for configuring or controlling the operation of the interface, and a *status register* indicates information about the state or condition of the interface. An I/O interface can include any number and combination of these registers, from a simple I/O interface consisting of a single data register to a complex I/O interface with several data, control and status registers. Some registers are subdivided into individual bits that serve as status, control or data bits individually.

To communicate with the I/O interface, the CPU must have the ability to read and write I/O interface registers. Just as with memory, each of these I/O registers is assigned a unique identifier through which it is addressed by the CPU. These I/O addresses form an address space that can either be part of the CPU's memory address space or be separate from it. When memory and I/O share the same address space, the CPU is said to use *memory-mapped I/O*; otherwise, it is said to use *separate I/O*. Since memory-mapped I/O interface registers are mapped into the CPU's memory address space, these registers are manipulated by software in the same way as memory bytes, using existing CPU instructions and addressing modes. This allows much flexibility when working with I/O registers, but reduces the amount of the address space that can be mapped to RAM or ROM. With separate I/O, special additional CPU instructions or addressing modes are needed to access I/O registers due to the separation of the two address spaces. The HCS08 CPU uses memory-mapped I/O.

An example of a fictional memory-mapped I/O interface is shown in Figure 1.2. As shown,

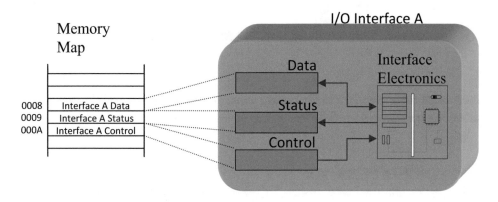

Figure 1.2: Illustration of a memory-mapped I/O interface.

the data register has been mapped into the CPU address space at address 0008_{16}, the status register at 0009_{16} and the data register at $000A_{16}$. The CPU can read the contents of the data register with a load from address 0008_{16}, for example, or configure a property of the interface by manipulating bits at address $000A_{16}$.

Although memory-mapped I/O registers are accessed as though they were memory bytes, they generally do not behave like RAM or ROM. One reason is that the data in an interface register can be changed outside of program control, making its contents volatile (successive loads from the register can return different values). Another reason is that not all I/O registers have both read and write capability. For example, a data register in an input interface might be read-only by software and a store to the associated memory address would have no effect on the contents of the register.

Figure 1.3 shows the memory maps of the MC9S08QG8 and MC9S08QG4 microcontrollers. For both devices, some I/O port registers are mapped in the direct page (addresses $0000–$005F) and some are mapped in the high page ($1800–$184F). Direct page registers tend to be registers associated with I/O ports that are accessed more frequently by software, while high page registers tend to be registers that are for system configuration and likely to be accessed only once after a

$0000 $005F	Memory-Mapped I/O Registers		$0000 $005F	Memory-Mapped I/O Registers
$0060 $025F	RAM (512B)		$0060 $015F	RAM (256B)
	unmapped			unmapped
$1800 $184F	Memory-Mapped I/O Registers		$1800 $184F	Memory-Mapped I/O Registers
	unmapped			unmapped
$E000 $FFFF	Flash (8KiB)		$F000 $FFFF	Flash (4KiB)
	MC9S08QG8			MC9S08QG4

Figure 1.3: Memory maps of the MC9S08QG8 and MC9S08QG4 microcontrollers.

system reset. Since there are only 256 direct page locations, mapping infrequently used I/O registers to the extended page allows more RAM to be mapped to the direct page.

1.1.3   I/O SYNCHRONIZATION

During program execution, software accesses I/O ports to exchange data with the peripheral devices connected to them. The times at which software accesses the ports will rarely correspond to the times when the I/O device connected to the port is ready to accept or provide data. For example, consider a keypad connected to a microcomputer system via an input port. When a program is ready to accept user input, it reads the port associated with the keypad to determine which key is pressed. If this read does not occur in the interval during which the user is pressing the key, the key press will not be detected. In general, a mechanism is needed to coordinate software accesses to I/O ports with the timing characteristics of the device connected to the port. This coordination is called I/O synchronization. In general, there are three forms of I/O synchronization: polling, real-time synchronization and interrupts.

Polling is the simplest of the three types of I/O synchronization. When performing polled I/O, software repeatedly checks (polls) the status of the device connected to the port to determine if the input or output operation can be performed. Thus, a fundamental requirement for using polled-I/O is the ability to query the state of the device. If a device contains a status register, this status register holds information on the state of the interface or the device; this state is often used for polling. In addition, it is sometimes possible to poll a data register for a particular value that indicates the state of the device. For example, it is possible to determine the state of a push-button switch connected to a general-purpose I/O pin by reading the logic value reflected in the data register associated with the port.

The primary advantage of polling is its simplicity. The main disadvantage of polling is that the CPU is idle while polling; that is, it cannot execute other useful instructions unless they can be included within the polling loop.

Delay synchronization uses software delays to match program timing to I/O timing. This form of synchronization is useful if I/O operations have predictable timing characteristics. For example, suppose a peripheral device can accept data at a maximum rate of 1 byte per millisecond. Since the CPU can write data to the port at a much higher rate that this, synchronization can be achieved by inserting a 1 millisecond delay between subsequent writes to the port. In some cases, the CPU requests data from an I/O device and the device requires a finite amount of time to process the request. An example is an analog to digital convertor, which requires several clock cycles to compute a digital approximation of the analog voltage. In these cases, a fixed delay can be inserted between the request and the load of data from the device, allowing sufficient time for the I/O operation to complete. Delay synchronization is useful when there is no status to poll and when I/O operations have predictable timing characteristics.

Delay synchronization is slightly more complex than polling because of the need to generate accurate software delays, typically using timed software loops. If polling is possible, it is generally easier and more efficient to do. In addition, because the CPU is idle while executing software delay loops, delay synchronization is no more efficient than polling.

Interrupt Synchronization uses the CPU hardware interrupt mechanism to interrupt running programs when an I/O event has occurred. This synchronization mechanism is generally the most efficient since the CPU is interrupted only when I/O processing needs to occur. Hardware support, in the form of integration with the CPU interrupt mechanism, is required as is the need to create an interrupt service routine, which is a special subroutine used to process the I/O. Interrupt synchronization is covered in more detail in Section 1.4.

1.1.4   DEVICE DRIVERS

The CPU interacts with I/O interfaces through interface registers, which provide for low-level control of the interface and access to data. When designing programs that access I/O, the embedded systems programmer is more concerned with high-level operations; the low-level interface details are of little concern. For example, the programmer is concerned with what key is being pressed on a keypad and not necessarily with how the keypad is interfaced or what steps are required to obtain the key value from the I/O interface. Having to perform low-level access to I/O interfaces overly complicates the work the programmer must do to access and control the peripheral on the interface. By writing a set of subroutines to manipulate the I/O interface and peripheral, the programmer can subsequently access the peripheral at a higher level of abstraction, without managing the underlying details of how the interface actually works. This set of subroutines is called a *device driver*.

Device drivers have all of the benefits of modular program development associated with subroutines in general, including code reuse and ease of program maintenance. The device driver provides high-level abstraction that speeds program development because it separates the requirements of the application from the specifics of the hardware. This abstraction also allows the same program to work with different types of interfaces and peripherals, as long as the drivers for each interface/peripheral combination provide the same functional abstraction to the program. This abstraction also allows a device of one type to look like another type to software (for example, connect a digital camera to your PC and it looks like a removable hard drive). This is how modern operating systems seamlessly integrate the hardware of different manufacturers and allow different types of devices to be manipulated in similar ways.

Device driver subroutines that use polled I/O can be either *blocking* or *non-blocking*. A blocking subroutine does not return back to the calling program until the I/O operation performed by the subroutine is complete. In a non-blocking subroutine, the subroutine returns an error code if the I/O operation cannot complete, allowing software to continue if possible.

1.2   A MC9S08QG4/8 SKELETON PROGRAM

In addition to device drivers to access I/O and software to manipulate data, a complete program for an embedded microcontroller must include some start-up code to configure the microcontroller after reset as well as perform basic system management required to keep the microcontroller running. Unlike a general-purpose computer, an embedded microcomputer does not always have an operating system that performs these startup and system management functions. The minimal set of such functions necessary to start up and keep a basic MC9S08QG4/8 system running are described in this section. These include configuring system registers, managing the watchdog timer, creating a simple interrupt service routine and programming the interrupt vector table.

1.2.1   SYSTEM CONFIGURATION REGISTERS

The MC9S08QG4/8 contains two system options registers, SOPT1 and SOPT2, shown in Figure 1.4. These high page registers are write-once registers, meaning that subsequent writes to them have no effect. This prevents erroneous software from altering critical system configuration parameters, which could cause the system to become unstable. This also means that any MC9S08QG4/8 software should always write to these registers once after reset, even if the default values are being used, to prevent such errors from occurring. The default values of the registers after reset are shown below each register in the figure. Because these are I/O registers and not memory, the values returned by a load do not necessarily reflect those last written. When the read and write behavior of a register is different, the read behavior is shown on the top half of each bit and the write behavior on the bottom. For example, software will always read zero for bits 2 through 6 of SOPT2, independent of the values written to these bits.

The configuration bits COPE, COPT, and COPCLKS control the configuration of the computer operating properly (COP) watchdog, which is described in detail in Section 1.2.2. STOPE enables or disables STOP mode. STOP mode is a low power mode that the microcontroller enters upon execution of a STOP instruction. When disabled, execution of a STOP instruction will instead cause an illegal opcode reset.

SOPT1: Systems Option Register 1 (memory-mapped at address $1802)

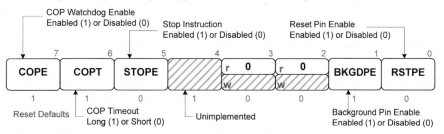

SOPT2: Systems Option Register 2 (memory-mapped at address $1803)

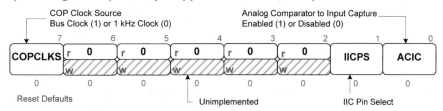

Figure 1.4: System options registers.

BKGDPE is the background debug pin enable bit. Background debug mode allows software tools to analyze microcontroller operation during software development. Using background debug mode, memory and register contents can be inspected and modified and the CPU can be controlled to implement single-stepping and other debug functions. A single wire debug connection on PTA4 (pin 2) is used by development tools to access and control the microcontroller. When the system in under development, BKGDPE needs to be enabled (BKGDPE=1) for such debugging. When the system is ready for production, if pin 2 is to be used as a general-purpose I/O pin in the final system, the BKGDPE must be disabled.

Reset pin enable (RSTPE) enables the active-low reset function on microcontroller pin 1. When enabled, an active low pulse on pin 1 forces a CPU reset. When RSTPE is disabled, pin 1 can be used as an I/O pin.

IICPS is the pin select configuration bit for the inter-integrated circuit (IIC) module. When the IIC module is enabled, it can be configured to use either pins 6 and 7 or pins 13 and 14 (IIC uses two pins called SDA and SCL). When the IIC module is not being used, the value of IICPS is irrelevant. When set, ACIC (Analog Comparator to Input Capture) enable connects the output of analog comparator module to the input of the timer/pulse-width modulator.

1.2.2 COMPUTER OPERATING PROPERLY (COP) WATCHDOG

The computer operating properly (COP) watchdog is a circuit intended to reset the CPU in the event of a software error. The COP watchdog consists of a free-running counter that is configured to count at a certain rate. If the counter reaches its terminal value, then a system reset is forced. In order to avoid a watchdog reset, software must reset the COP counter periodically.

The COP watchdog is configured via the system options registers, SOPT1 and SOPT2, as described in Section 1.2.1. The COPE (COP Enable) bit of SOPT1 selects whether the COP is enabled or disabled. When disabled, a COP watchdog reset never occurs. When enabled, the COPCLKS bit (bit 7 of SOPT2) selects the clock source to the COP counter controlling the counting rate. This rate can be configured to be the bus clock frequency (COPCLKS=1), which is 4 MHz by default out of reset, or a separate 1 kHz internal clock (COPCLKS=0). The COPT (COP Timeout) bit in SOPT1 selects a short or long timeout period. When the 1 kHz clock is selected as the COP clock source, a short timeout is defined as 32 periods of the clock (32 ms) and a long timeout is defined as 256 clock periods (256 ms). When the bus clock is selected as the COP clock source, a short timeout is defined as 2^{13} periods (8.192 ms at 4 MHz) and a long timeout 2^{18} periods (0.262 s at 4 MHz). The timeout period is selected depending on system design requirements.

Resetting the COP counter consists of writing any value to the SRS register, which is memory mapped to location $1800. Embedded software is often structured to have a main loop that repeats forever. In such applications, the reset of the COP counter can be placed somewhere in this main loop such that the counter will be reset once per iteration. As long as the execution time through the loop is shorter than the COP timeout period, this approach will prevent a COP reset unless there is a software error such as an infinite loop. If the code in the main loop includes blocking I/O operations, care should be taken to prevent the COP reset if the blocking time exceeds the COP timeout period. This is usually accomplished by resetting the COP counter in the I/O polling or delay loop.

1.2.3   INTERRUPT VECTOR TABLE

Resetting the microcontroller is a way of initializing it to a known operating state. Upon reset, I/O registers default to their reset states (usually disabled) and the I bit in the CCR is set to block maskable interrupts, allowing the programmer to configure the system to a safe state before interrupts can occur. An interrupt is an asynchronous event that can occur at almost any time while the microcontroller is running. When either a reset or interrupt occurs, the CPU needs an address in memory (a vector) at which it should start executing. In the case of a reset, the vector is the address of the start of the system code; in the case of an interrupt, the vector is the address of the start of the interrupt service routine that will deal with the request.

The CPU maintains a table of reset and interrupt vectors that must be correctly initialized at the time the microcontroller is programmed. Table 1.1 lists each vector location and its associated interrupt or reset source. Each vector requires two bytes of storage to hold an address. Note that the addresses in the table are not contiguous; for example, there are no vectors located from $FFD2 to $FFD5. These locations of the vector table are reserved for future use.

Table 1.1: Interrupt/Reset Vector Locations of the MC9S08QG4/8.

Vector Address	Interrupt/Reset Source
$FFD0:FFD1	Real Time Interrupt (RTI)
$FFD6:FFD7	Analog Comparator (ACMP)
$FFD8:FFD9	Analog to Digital Converter (ADC)
$FFDA:FFDB	Keyboard Interrupt (KBI)
$FFDC:FFDD	Inter-Integrated Circuit (IIC)
$FFDE:FFDF	Serial Communications Interface (SCI) Transmit
$FFE0:FFE1	Serial Communications Interface (SCI) Receive
$FFE2:FFE3	Serial Communications Interface (SCI) Error
$FFE4:FFE5	Serial Peripheral Interface (SPI)
$FFE6:FFE7	Modulo Timer (MTIM) Overflow
$FFF0:FFF1	Timer/Pulse Width Modulator (TPM) Overflow
$FFF2:FFF3	Timer/Pulse Width Modulator (TPM) Channel 1
$FFF4:FFF5	Timer/Pulse Width Modulator (TPM) Channel 0
$FFF8:FFF9	Low Voltage Detect Reset
$FFFA:FFFB	Interrupt Request Pin (IRQ) Reset
$FFFC:FFFD	Software Interrupt (SWI)
$FFFE:FFFF	CPU Reset

1.2.4 HCS08 MODES OF OPERATION

The HCS08 CPU has 4 primary modes of operation: run, wait, stop and active background. The run mode is the primary mode of operation in which the CPU sequences instructions and peripherals operate normally. Run mode is entered upon power-on reset when the BKGD pin is high. In run mode, instruction sequencing begins at the address in the reset vector. Active background mode is a debugging mode that is useful when developing and testing new software. In this mode, the CPU is in a suspended state and debugging commands are accepted over the BKGD pin. These debugging commands allow for inspection and modification of registers or memory locations and for single stepping instructions. Active background mode can be entered in several ways, including the CPU BKGD instruction and holding the BKGD pin low during power-on reset.

Wait mode is entered upon execution of the WAIT instruction. In this mode, the CPU ceases instruction sequencing and is not clocked, lowering power consumption. In wait mode, peripherals operate normally and any interrupt request automatically wakes-up the CPU and places it back into run mode. Since an interrupt is required to wake the CPU, the CPU always resumes instruction sequencing with the interrupt service routine corresponding to the source of the interrupt that woke the CPU. After the service routine completes, instruction sequencing resumes after the WAIT instruction that put the CPU in wait mode. The WAIT instruction always clears the I mask in the

CCR to ensure that interrupts are enabled. At least one interrupt source must be enabled to wake the CPU or it will wait indefinitely.

There are three levels of stop mode provided in the CPU. In all levels, the CPU and most peripheral devices are shut down to minimize power consumption. Stop mode is entered by executing the STOP instruction and the stop level is determined by system configuration register settings. Stop level 1 provides maximum power savings and requires a power-on reset to restart the system. Stop level 2 is an intermediate level that maintains the state of RAM and I/O pins. A real-time interrupt can restart the system from stop level 2. Stop level 3 (the default level) maintains the same system state as stop level two in addition to the CPU register contents.

1.2.5   PROGRAM SKELETON FOR THE MC9S08QG4/8

Code Listing 1.1 is a complete *program skeleton* for the MC9S08QG4/8. A program skeleton is a template that outlines program structure and contains code that is common to all programs. Lines 4 through 6 define the memory locations within the memory map. On all members of the microcontroller family, the start of RAM is at address $0060. This is defined using an equate pseudo-op on line 4. The RAM size and start of flash ROM, however, differ for various microcontroller family members and the equates for RAMSIZE and FLASHSTART on lines 5 and 6 must be programmed accordingly. Lines 7 through 13 contain equate pseudo-ops that define the memory-map locations of individual I/O registers described previously.

Lines 17 through 24 contain system configuration constants that are use by the initialization code to configure the microcontroller. Constant CLOCKFREQ is a 2-bit value used to establish the operating frequency of the microcontroller clock (bus clock). These two bits are written to the ICSSC2 register in the system initialization code. COPSET is used to select the desired COP setting; a value of 0 disables the COP, while values of 2 and 3 enable it and select a short or long timeout period, respectively. STOPEN, when set to 1, enables the STOP instruction, which allows programs to direct the CPU to enter stop mode. BKGDPEN, when set, enables the BKGD function on pin 2; this must be enabled during system development when debugging, but disabled if other functions on pin 2 are to be used (PTA4 or analog comparator output ACMPO). RSTPEN is used to enable the active low reset function on pin 1; this must be cleared if the reset function is to be disabled or if the IRQ or PTA5 functions are needed on this pin. The constant SOPT1VAL combines the COPSET, STOPEN, BKGDPEN and RSTPEN constants to form the value that needs to be written to system options register 1. Similarly, SOPT2VAL is the value that is written to system options register 2.

Line 28 contains the ORG pseudo-op for RAM. All variable definitions intended for RAM should follow this ORG. Line 33 contains a similar pseudo-op defining the location of flash memory. The first instructions following this ORG on line 33 are system initialization instructions that should not be changed without careful consideration. Lines 34 through 37 program the write-once system options registers.

The onboard clock generator is not very accurate and can be off by a considerable extent. Software can tweak the clock frequency by writing a trim value to the ICSTRIM system configuration

```
 1  ;--------------------------------------------------------------------
 2  ; Memory Map Definitions
 3  ;--------------------------------------------------------------------
 4  RAMSTART    equ   $0060
 5  RAMSIZE     equ   256    ; this is 256 on QG4, 512 on QG8
 6  FLASHSTART  equ   $f000  ; this is $F000 on QG4, $E000 on QG8
 7  WATCHDOG    equ   $1800  ; location of watchdog's food dish
 8  VECTOR_TBL  equ   $ffd0  ; start of interrupt vector table
 9  SOPT1       equ   $1802  ; system options registers
10  SOPT2       equ   $1803
11  ICSTRIM     equ   $003A  ; clock trim register
12  ICSSC2      equ   $0039  ; clock status and control register
13  ICSTRIMVAL  equ   $FFAF  ; factory programmed clock trim value
14  ;--------------------------------------------------------------------
15  ; System Configuration OPtions
16  ;--------------------------------------------------------------------
17  CLOCKFREQ   equ   0      ; 0 - 8MHz; 1 - 4MHz; 2 - 2MHz; 3 - 1MHz
18  COPSET      equ   3      ; 0- disable COP, 2- enable with short timeout
19                          ; 3- enable with long timeout
20  STOPEN      equ   1      ; 0- disable stop instruction, 1- enable
21  BKGDPEN     equ   1      ; 0- disable BKGD function, 1- enable
22  RSTPEN      equ   0      ; 0- disable RST pin, 1- enable
23  SOPT1VAL    equ   (COPSET<<6| STOPEN<< 4 | BKGDPEN<<1 | RSTPEN )
24  SOPT2VAL    equ   $00    ;COP clk is 1kHz, IICPS and ACIC at defaults
25  ;--------------------------------------------------------------------
26  ; Variable data section: ORGed to start of RAM
27  ;--------------------------------------------------------------------
28              org   RAMSTART
29  ExampleVar: ds.b  1
30  ;--------------------------------------------------------------------
31  ; Code Section: ORGed to start of flash
32  ;--------------------------------------------------------------------
33              org   FLASHSTART
34  MAIN:       lda   #SOPT1VAL         ;configure system options
35              sta   SOPT1
36              lda   #SOPT2VAL
37              sta   SOPT2
38              lda   ICSTRIMVAL        ;load factory programmed trim value
39              sta   ICSTRIM           ;write to clock trim register
40              mov   #(CLOCKFREQ<<6),ICSSC2 ;set internal clock frequency
41              ldhx  #(RAMSTART+RAMSIZE) ;initialize the stack pointer
42              txs
```

Code Listing 1.1: Skeleton Program for the MC9S08GQ4/8 (*Continues*).

```
MAININIT:   nop                  ; program initialization
            cli                  ; enable interrupts after system init
MAINLOOP:   nop                  ; main loop body

FEEDTHEDOG: sta    WATCHDOG      ; reset the watchdog
            bra    MAINLOOP
;-------------------------------------------------------------------------
; Constant Section: Not ORGed, follows code section in flash
;-------------------------------------------------------------------------
NULL        dc.b   0
;-------------------------------------------------------------------------
; Dummy ISR to catch spurious interrupts
;-------------------------------------------------------------------------
DUMMY_ISR:  bra    DUMMY_ISR     ;stay here to force COP reset
;-------------------------------------------------------------------------
;                  Interrupt Vector Table
;-------------------------------------------------------------------------
            org    VECTOR_TBL
            dc.w   DUMMY_ISR ;  $FFD0:FFD1 RTI
            ds.w   2         ;  $FFD2:FFD5 ***Reserved***
            dc.w   DUMMY_ISR ;  $FFD6:FFD7 ACMP
            dc.w   DUMMY_ISR ;  $FFD8:FFD9 ADC Conversion
            dc.w   DUMMY_ISR ;  $FFDA:FFDB KBI Interrupt
            dc.w   DUMMY_ISR ;  $FFDC:FFDD IIC
            dc.w   DUMMY_ISR ;  $FFDE:FFDF SCI Transmit
            dc.w   DUMMY_ISR ;  $FFE0:FFE1 SCI Receive
            dc.w   DUMMY_ISR ;  $FFE2:FFE3 SCI Error
            dc.w   DUMMY_ISR ;  $FFE4:FFE5 SPI
            dc.w   DUMMY_ISR ;  $FFE6:FFE7 MTIM Overflow
            ds.w   4         ;  $FFE8:FFEF ***Reserved***
            dc.w   DUMMY_ISR ;  $FFF0:FFF1 TPM Overflow
            dc.w   DUMMY_ISR ;  $FFF2:FFF3 TPM Channel 1
            dc.w   DUMMY_ISR ;  $FFF4:FFF5 TPM Channel 0
            ds.w   1         ;  $FFF6:FFF7 ***Reserved***
            dc.w   DUMMY_ISR ;  $FFF8:FFF9 Low Voltage Detect
            dc.w   DUMMY_ISR ;  $FFFA:FFFB IRQ
            dc.w   DUMMY_ISR ;  $FFFC:FFFD SWI
            dc.w   MAIN      ;  $FFFE:FFFF Reset
```

Code Listing 1.1: (*Continued*) Skeleton Program for the MC9S08GQ4/8.

register. Freescale has defined flash memory location $FFAF as the predefined location of the trim value necessary to trim the clock to 4 MHz out of reset. Each microcontroller has a trim value stored in memory at the factory, and many development tools compute and reprogram this value each time flash is programmed. Lines 38 and 39 use this value to trim the system clock. Subsequently, the MOV instruction on line 40 sets the clock divisor bits in the system clock status and control register to define the desired bus clock frequency.

Lines 41 and 42 initialize the stack pointer to point to the last byte in RAM. The NOP instruction on line 43 is a placeholder for the main initialization code needed to set up the microcontroller peripherals and global variables for a specific application. This usually involves calling driver initialization routines to initialize peripherals and setting global variables to their initial values. Line 44 enables interrupts by clearing the I mask bit in the CCR. This must be done after driver initialization to ensure that no interrupts are triggered until peripherals are properly configured.

The NOP on line 45 is another placeholder. It should be replaced with the body of the main program loop that is repeated for as long as the microcontroller is running. Once per iteration of this main loop, on line 47, the COP watchdog counter is reset. If the execution time of the main loop could exceed the configured watchdog timeout period, the watchdog may need to be at other locations within the main loop body. Following this is the branch back to the beginning of the main program loop that causes the loop to repeat forever.

Program constants, drivers and subroutines follow immediately after the main program loop and do not require an ORG pseudo-op. A default interrupt service routine, DUMMY_ISR, is provided to initialize interrupt vectors that are not in use. This helps to catch spurious interrupts caused by misconfiguration of a peripheral. The BRA DUMMY_ISR instruction keeps the CPU at the ISR until the COP resets the processor, allowing the system to safely recover from the error. Good programming will ensure that this never happens.

Line 60 begins the definition of the interrupt vector table. All interrupt vectors not assigned to a specific service routine are initialized with a DC.W pseudo-op that points the vector to DUMMY_ISR. The CPU reset vector, at $FFFE:$FFFF, represents the address of the first instruction executed after a CPU reset. This is programmed to point to label MAIN, which corresponds to the line of the first instruction of the program.

1.3   GENERAL-PURPOSE DIGITAL I/O

General purpose digital I/O, the simplest form of I/O in an embedded system, allows software to directly control the logic levels on microcontroller pins. In addition to introducing general purpose digital I/O concepts, this section describes the electrical characteristics of the MC9S08QG microprocessor family.

A general-purpose input port uses microcontroller pins as digital inputs. A load from a data register associated with the input port will return the logic values on the pins. A general-purpose output port uses the pins as digital outputs, controlled via a store to the data register associated with interface. A general-purpose input/output port uses the pins as either digital inputs or outputs, configurable through a control register called a *data direction register*.

1.3.1   GENERAL PURPOSE I/O ON THE MC9S08QG4/8

MC9S08QG4/8 pins labeled PTBi and PTAi correspond to pins that can be used as general-purpose I/O (GPIO) pins. These are organized into two ports: Port A and Port B. Port A is associated with 6 pins, labeled PTA0–PTA5, and port B is associated with 8 pins, PTB0–PTB7. The signals for port B are not available on an 8-bit package. Each of these GPIO pins can be independently configured as input or output, except input only PTA5 and output only PTA4.

Figure 1.5 illustrates the format of the Port A and Port B data and control registers. Pin PTBx

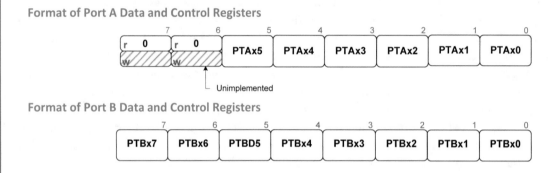

Figure 1.5: General purpose I/O port register formats.

can be individually configured as input or output via bit *x* in the I/O configuration register called the Port B *Data Direction Register* (PTBDD), which is memory-mapped to address $0003. When PTBDD bit *x* is a '1', the pin is an output driven by the logic value in bit *x* of the Port B *Data Register* (PTBD), which is memory-mapped to $0002. When PTBDD bit *x* is 0, pin PTBx is configured as an input. A read from bit x of PTBD returns the current logic value on the input pin. Table 1.2 summarizes the configuration options for port B as well as the value returned on a read from the port B data register. Note that a read from the port B data direction register always returns the current pin direction. The Port A operation is similar. Port A data (PTAD) is memory mapped to $0000, while its direction

Table 1.2: Summary of configuration settings for PTB.

PTBDDx	PTBDx	Pin Configuration	Read from Port B Data Register (PTBD)
0	-	Input	Returns logic value on input pin
1	0	Output driving low	Returns 0
1	1	Output driving high	Returns 1

register (PTADD) is mapped to $0001. Note that pin PTA4 is an output-only pin and pin PTA5 is an input only pin, and there is no pin PTA6 or PTA7.

Table 1.3 summarizes the configuration registers associate with GPIO ports A and B. Note

Table 1.3: Summary of GPIO port data and configuration registers.			
Register	**Address**	**Reset Default**	**Description**
Port A Data (PTAD)	$0000	$00	Port A I/O data register.
Port A Data Direction (PTADD)	$0001	$00 (input)	Port A I/O direction control. PTA4 is always output (clearing PTADD4 has no effect); PTA5 is always input (setting PTADD5 has no effect).
Port A Pull-up Enable (PTAPE)	$1840	$00 (disabled)	Port A pull-up resistor enable. PTAPEx=1 enables pull-up resistor on pin PTAx.
Port A Slew Rate Control (PTASE)	$1841	$3F (enabled)	Port A slew rate control; PTASEx=1 enables slew rate limiting on pin PTAx.
Port A Drive Strength (PTADS)	$1842	$00 (low drive)	High drive strength control; high drive strength enabled on pin PTAx when PTADSx=1.
Port B Data (PTBD)	$0002	$00	Port B I/O data register.
Port B Data Direction (PTBDD)	$0003	$00 (input)	Port B I/O direction control.
Port B Pull-up Enable (PTBPE)	$1844	$00 (disabled)	Port B pull-up resistor enable. PTBPEx=1 enables pull-up resistor on pin PTBx.
Port B Slew Rate Control (PTBSE)	$1845	$FF (enabled)	Port B slew rate control; PTBSEx=1 enables slew rate limiting on pin PTBx.
Port B Drive Strength (PTBDS)	$1846	$00 (low drive)	High drive strength control; high drive strength enabled on pin PTBx when PTBDSx=1.

that when a pin is configured as input, a write to port data register has no effect on the pin, but the value written is stored in the data register and will become the value driving the pin if its direction is changed from input to output. This allows the logic value on the pin to be correctly configured before the pin is enabled as an output, preventing glitches.

An internal pull-up resistor for each GPIO pin can be selectively enabled in an I/O control register called a *Pull-Up Enable* register. This internal pull-up eliminates the need to include an external resistor on the pin when one is required. Pull-ups are automatically disabled, regardless of the pull-up register setting, when the pin is configured as output.

The GPIO ports on the MC9S08QG4/8 microprocessors also have configuration registers that control drive strength and slew rate. Slew rate defines how quickly an output can change logic levels. Because fast switching outputs can cause higher electromagnetic interference (EMI), enabling slew rate control on a pin reduces how quickly it can switch, thereby reducing EMI. Drive strength control defines how much current the pin can source or sink. Under low drive strength, an output pin can source or sink a current of up to approximately 5-10 mA; when high drive strength is enabled, the maximum current increases to approximately 15-20 mA.

Example 1.1.   Write the instructions necessary to configure all 8 pins of Port B as input pins, with internal pull-up registers enabled.

Solution: Since the entire port is being used, configuration data can be directly written to the data direction register and pull-up enable register. Since PTBPE is an extended address, MOV cannot be used.

Answer:

```
1   PTBD    equ   $02          ;port B data register
2   PTBDD   equ   $03          ;port B direction control register
3   PTBPE   equ   $1844        ;port B pull-up control register
4   INITSW: lda   #$FF
5           sta   PTBPE        ;enable pull-ups on all Port B pins
6           mov   #$00,PTBDD   ;configure all port B pins as input pins
```

Example 1.2.   Write the instruction(s) necessary to configure pin PTA1 as an output pin with low drive strength and slew rate control on.

Solution: Since a single pin is being configured, it is required to modify only one bit of each control registers. BCLR and BSET are useful for registers in the direct page; general masking must be used for high page registers. To make PTA1 an output pin, its bit in PTADD must be set. PTASE1=1 to turn on slew rate control, and PTADS1=0 for low drive strength.

Answer:

```
1   PTAD    equ   $0000        ;port B data register
2   PTADD   equ   $0001        ;PTB direction control register
3   PTAPE   equ   $1840        ;PTB pull-up control register
4   PTASE   equ   $1841        ;port A slew control enable
5   PTADS   equ   $1842        ;port A drive strength control
6
7   INITSW: bclr  1,PTAD       ;write initial 1 to Port A data register
8           lda   PTADS
9           and   #%11111101
10          sta   PTADS        ;PTA1 drive strength low
11          lda   PTASE
12          ora   #%00000010
13          sta   PTASE        ;slew rate control on for PTA1
14          bset  1,PTADD      ;configure PTA1 as output
```

Example 1.3. Write the instructions necessary to configure pins PTB3, PTB2, PTB1, and PTB0 as input pins with pull-ups enabled, without changing the configuration of the other PTB pins. Solution:Since the entire Port is not being configured, data cannot be directly written to the data direction register or pull-up enable register. Thus, masking operations should be used.

Solution: Since the entire Port is not being configured, data cannot be directly written to the data direction register or pull-up enable register. Thus, masking operations should be used.

Answer:

```
1    PTBD      equ   $0002        ;port B data register
2    PTBDD     equ   $0003        ;PTB direction control register
3    PTBPE     equ   $1844        ;PTB pull-up control register
4    ONE_MASK  equ   %00001111    ;ONE mask of pins being configured
5    ZERO_MASK equ   %11110000    ;ZERO mask of pins being configured
6
7    INITSW:   lda   PTBPE        ;load current PTBPUE configuration
8              ora   #ONE_MASK    ;set bits 3-0 (enable pull-ups)
9              sta   PTBPE        ;write back modified configuration
10             lda   PTBDD        ;load current DDR configuration
11             and   #ZERO_MASK   ;clear bits 3-0 (configure as inputs)
12             sta   PTBDD        ;write back modified DDR configuration
```

1.3.2 ELECTRICAL SPECIFICATIONS

The electrical specifications, found in the *MC9S08QG8/MC9S08QG4 Data Sheet*, define the electrical, timing and temperature conditions under which the device can be safely operated. When integrating hardware into the microcontroller-based system, these specifications must be considered to ensure proper operation of the device. The device specifications include *Absolute Maximum Ratings* that define the limits to which the device can be exposed without causing permanent damage to it. The *Functional Operating Range* defines the valid voltage, current, timing and temperature ranges under which the device will function correctly. The functional operating range is always within the absolute maximum ratings. While it is possible to subject the device to conditions outside the functional range, but within the absolute maximum ratings, the device is not guaranteed to operate properly.

Table 1.4 lists the absolute maximum ratings for the MC9S08QG4/8, obtained from the device data sheet. All voltages are relative to the supply voltage ground, V_{SS}, which is 0V or ground by definition. The supply voltage V_{DD} must not be allowed to exceed +3.8V or −0.3V V_{SS}. The *Digital Input Voltage* characteristic, V_{IN}, defines the allowable voltage range on all pins (other than V_{DD} and V_{SS}). The voltage on any pin must not exceed $V_{DD} + 0.3V$ or drop lower than $V_{SS} - 0.3V$. This is especially important to consider when interfacing external analog circuits connected to these pins and in mixed-voltage digital systems. If the voltage on a data pin should exceed V_{IN}, current will flow into (or out of) the pin and can damage the device; an appropriate current limiting circuit

Table 1.4: Absolute maximum ratings for the MC9S08QG4/8.

Rating	Symbol	Value	Unit
Supply voltage	V_{DD}	−0.3 to +3.8	V
Maximum current into VDD	I_{DD}	120	mA
Digital input voltage	V_{IN}	−0.3 to VDD+0.3	V
Instantaneous maximum current per port pin	I_D	± 25	mA
Storage temperature range	T_{STG}	−55 to 150	°C

must be used to prevent this. The maximum power supply current is 120mA (parameter I_{DD}); this includes all current being sourced by microcontroller pins in addition to that used to operate the CPU and peripherals. The instantaneous maximum current, I_D, for data pins is ± 25 mA.

Table 1.5 lists selected electrical characteristics in the functional operating range for the MC9S08QG4/8 devices. The functional operating range for the MC9S08QG4/8 specifies that

Table 1.5: Selected electrical characteristics for the MC9S08QG4/8.

Parameter	Symbol	Minimum Value	Maximum Value	Unit
Supply Voltage	V_{DD}	1.8	3.6	V
Input High Voltage	V_{IH}	$.7{\times}V_{DD}$	-	V
Input Low Voltage	V_{IL}	-	$.3{\times}V_{DD}$	V
Digital Input Hysteresis	V_{HYS}	$0.06{\times}V_{DD}$	-	V
Input Leakage Current (per pin)	$|I_{IN}|$	-	1.0	µA
Internal Pull-up Resistance	R_{PU}	17.5	52.5	kΩ
Output High Voltage	V_{OH}	Depends on IOH (see text)		V
Output High Current (sum, all pins)	$|I_{OHT}|$	-	60	mA
Output Low Voltage	V_{OL}	Depends on IOH (see text)		V
DC Injection Current (per pin)	I_{IC}	-0.2	0.2	mA
DC Injection Current (all pins)	I_{IC}	-5	5	mA

V_{DD} can be in the range from 1.8V to 3.6V. Many of the electrical specifications for the device depend on the supply voltage, temperature, and other operating conditions, often in a nonlinear way. The values listed in Table 1.5 are typical values for moderate operating conditions. If more precise values are required, the engineer needs to consult the device datasheet or perform laboratory measurements.

For digital input pins, the values of interest include the input high and low voltages, V_{IH} and V_{IL}, input leakage current I_{IN}, input hysteresis V_{HYS}, and the internal pull-up resistance R_{PU}. The values listed are typical and may be different at the extremes of the functional operating range. V_{IH} represents the range of voltages on an input pin that are interpreted as logic high. This range is defined by $.7V_{DD}$ on the low end and is limited by the maximum operating value for the device

(V_{DD} + 0.3V) on the high end. Likewise, the range of voltages that will be interpreted as logic low inputs is bounded by V_{SS} (0V) on the low end and .35V_{DD} on the high end. The input leakage current is the current that flows into and input pin. Though small, this current must be taken into account if large pull-up or pull-down resistors are used on inputs as it will cause a measurable voltage drop across the resistor. Usually, the maximum input leakage is the parameter of interest. Input hysteresis defines the minimum change in input voltage necessary to toggle the input logic level. For slowly changing inputs, a large hysteresis is desirable to prevent rapid input switching due to noise. Finally, each internal pull-up resistor has a value between 17.5 kOhm and 52.5 kOhm; the actual value depends on the value of the supply voltage; graphs of R_{PU} versus V_{DD} are provided in the data sheet.

A load on an output pin can be configured in a *sourcing configuration* or *sinking configuration*, as shown in Figure 1.6. In the figure, V_{LOAD} is the voltage drop across the load, V_{OUT} is the output

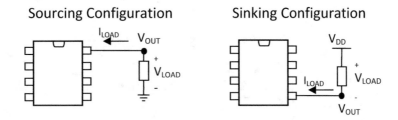

Figure 1.6: Illustration of sourcing and sinking output load configurations.

voltage at the pin (relative to ground) and I_{LOAD} is the load current, which by convention is defined as going into the pin.

For output pins, the parameters of interest are the output high and low voltages and the output current. The parameters V_{OH} and V_{OL} specify the voltage at the output pin when logic high and low values are output, respectively. In the sourcing configuration, the load is connected across the output pin and ground and $V_{LOAD} = V_{OUT}$. When the pin outputs a logic 1, $V_{LOAD} = V_{OH}$ and I_{LOAD} is negative (current flows out of the pin), thus the pin is said to be sourcing current. Because the pin is acting as a non-ideal voltage source, the actual value of V_{OH} depends on the output current I_{LOAD}. In the sourcing configuration, when the pin outputs a logic low, $V_{LOAD} = V_{OUT} = 0V$ and $I_{LOAD} = 0$ mA.

In the sinking configuration, the load is connected across the output pin and V_{DD} and $V_{LOAD} = V_{DD} - V_{OUT}$. When the pin outputs a logic 1, $V_{OUT} = V_{DD}$ and both V_{LOAD} and I_{LOAD} are zero. When the pin outputs a logic 0, $V_{LOAD} = V_{DD} - V_{OL}$ and I_{LOAD} is positive (current flows into the pin), thus the pin is said to be sinking current. V_{OL} depends on the amount of current the pin is sinking.

V_{OH} and V_{OL} both depend on I_{LOAD}, V_{DD}, device operating temperature and the drive strength configuration of the output pin. For the sinking configuration, V_{OL} increases from V_{SS} as I_{LOAD} increases; for the sourcing configuration, V_{OH} decreases (from V_{DD}) as I_{LOAD} increases. The data sheet provides graphs of the relationship for $V_{DD} = 3.0V$, but these can only be used as a starting point for approximation since actual values depend on many factors. A first-order approximation based on these graphs is shown in Figure 1.7. These approximations were made for $V_{DD} = 3.0V$ at

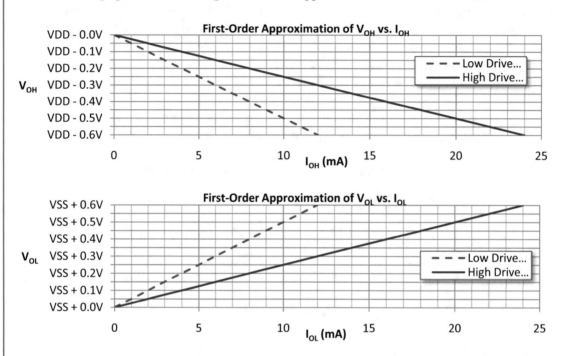

Figure 1.7: V_{OH} versus I_{OH} and V_{OL} versus I_{OL} curves for low and high drive strengths.

room temperature. The output high voltage characteristic shows the difference in voltage V_{OH} from its ideal value V_{DD} for the sourcing configuration, while the output low voltage characteristic shows the value of V_{OL} for the sinking configuration. Thus, both plots characterize the deviation from the ideal output voltage level. Clearly, as the load current increases in both configurations, the output voltage deviates increasingly from its ideal.

A model of this relationship is an ideal voltage source with a series resistor. The resulting Thevenin equivalent models of the output pin in sourcing and sinking configurations are shown Figure 1.8. R_{TH} is the Thevenin series resistance, which can be obtained from the slope of the $V - I$ curves above. From the figure, R_{TH} is approximately 50 Ohms for low drive strength and 25 Ohms for high drive strength. For either configuration, $V_{LOAD} = V_{DD} - I_{LOAD} \times R_{TH}$.

Note that there is a maximum combined I_{OH} parameter, $|I_{OHT}|$, of 60 mA and a maximum total $I_{OL}, |I_{OLT}|$, of 60 mA. This means that the sum of the load currents sourced for all output high pins must not exceed 60 mA. Similarly, the sum of all sinking load currents must not exceed 60 mA. The single pin limit is 25 mA, defined in the absolute maximum ratings.

Sourcing Pin Model

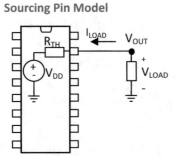

Sinking Pin Model

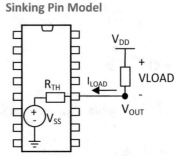

Figure 1.8: Thevenin equivalent circuit models for V_{OL} and V_{OH}.

The following are additional guidelines to consider when using the general-purpose I/O ports:

• Unused GPIO pins should never be left floating (that is, have no external circuit driving the pin to V_{DD} or V_{SS}). A floating input pin has an indeterminate voltage level that can cause excess internal current consumption. Unused GPIO pins should be configured as inputs with internal pull-ups enabled, or configured as output pins driving either a '1' or '0'.

• If external pull-down resisters are used, internal pull-up resisters must be disabled; otherwise, a voltage divider is formed and the input voltage may not represent a valid input high or input low.

• If glitches are a consideration for the circuit connected to an output pin, write the initial logic value desired to the data register before configuring the data direction register as output.

1.3.3   SWITCH INPUT INTERFACE

One of the simplest input sources is the mechanical switch input. A two-terminal mechanical switch is fashioned from two conductors that can be in one of two states: closed (in contact) or open (not in contact). While there are many types of switches available incorporating a variety of physical mechanisms to actuate the switch (temperature, motion, etc.), their integration into a microcomputer-based system is generally the same. Thus, while the focus of this section is on the integration of human-actuated switches, the same principles apply to other switch types.

Two types of simple mechanical switches are toggle switches and momentary switches. A toggle switch is set to the open or closed position and remains in that position until changed. A momentary switch has a normal state (open or closed) which is the state of the switch when there is no mechanical action; the switch is in the opposite state as long as the mechanical action lasts. The schematic symbols for common switches are shown in Figure 1.9. A human-actuated momentary switch is call a push-button switch.

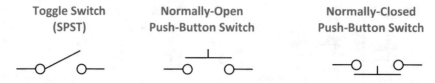

Figure 1.9: Schematic symbols for common mechanical switches.

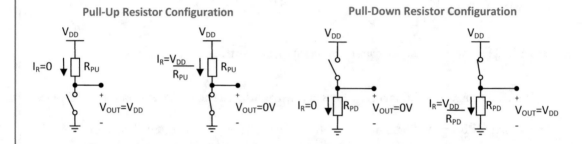

Figure 1.10: Switch input interface circuit configurations.

A switch by itself cannot provide a digital input source for general-purpose input. Instead, the switch must be included in a circuit that can convert the open and closed states of the switch into high and low logic levels. This is accomplished using a pull-up or pull-down resistor in series with the switch, as shown in Figure 1.10. When the switch is open in the pull-up resistor circuit, there is almost no current through the resistor (recall that there is the small leakage current, I_{IL}, drawn by the microcontroller input pin). Therefore, there is almost no voltage drop across the resistor and $V_{OUT} \approx V_{DD}$, resulting in a high logic output. When the switch is closed, V_{OUT} has a direct path to ground ($V_{OUT} = 0V$), resulting in a low-logic output; the purpose of the pull-up resistor in this case is to limit the current to $I = V_{DD}/R_{PU}$. The pull-down configuration works in a similar way, except that the logic levels are reversed. Generally, it is desirable to have as high a resistance for R_{PU} or R_{PD} as possible to limit wasted power when the switch is closed. The main limiting factor to using high resistance values is microcontroller input pin leakage current, which can cause a voltage drop across the resistor when the switch is open. This voltage drop must be limited to be well within the valid range for V_{IH} or V_{IL}, depending on the situation. This is addressed in the chapter problems.

When a momentary switch is used, the interface circuit is an active-high configuration if the logic level on the microcontroller pin is high when the switch is actuated. An active low configuration is one in which a logic low is on the microcontroller input pin when the switch is actuated. If the pull-up resistor switch configuration is used, then the internal pull-up resistor can be used instead of the external resistor; the resulting circuit functions in the same manner but eliminates one external circuit component. On the other hand, if the pull-down configuration is used, the pull-up resistor

must be disabled or a voltage divider will be formed, resulting in an intermediate voltage input when the switch is open, which can lead to erroneous circuit operation.

The software side of the interface is implemented with a device driver. The device driver can implement different functionality depending on the needs of the software that will use the switch input. In addition, the device driver can perform some processing of the input before it is returned to the requesting programs. This is illustrated in Code Listing 1.2. Lines 1-4 contain pseudo-ops used to define the memory-mapped I/O port locations as well as the pin number the interface is for, which minimizes code changes if the need to change the port or input pin used arises. These EQU pseudo-ops also make the code easier to read and understand and allow the code to be easily ported to other HCS08-based microcomputers, which might map the port registers to different addresses. INITSW is the driver initialization subroutine, which must be called from within the driver initialization section of the main program. INITSW uses masking operations to configure the pin as an input with internal pull-ups enabled. The OR-mask for the pin is formed by the assembler expression #(1<<SWPIN), which results in a mask with a single 1 in bit position SWPIN. This expression is computed by the assembler and does not result in additional instructions being generated.

```
1    PTBD      equ   $02              ;port B data register
2    PTBDD     equ   $03              ;PTB direction control register
3    PTBPE     equ   $1844            ;PTB pull-up control register
4    SWPIN     equ   3                ;defines the pin that switch is on
5    ;---------------------------------------------------------------------
6    INITSW:   psha                   ;callee save
7              bclr  SWPIN,PTBDD      ;only change setting for this pin
8              lda   PTBPE            ;enable internal pull-up resistor
9              ora   #(1<<SWPIN)      ;using an OR mask
10             sta   PTBPE            ;write back modified PTBPE value
11             pula                   ;callee restore
12             rts                    ;ISPUSHED
13   ;---------------------------------------------------------------------
14   ISPUSHED: brclr  SWPIN,PTBD,PRESSED ;detect if pressed
15             clc                    ;return C=0 if not
16             bra   END_ISP
17   PRESSED   sec                    ;else return C=1
18   END_ISP   rts
19   ;---------------------------------------------------------------------
20   WAITPRESS: sta   WATCHDOG        ;reset watchdog
21              brset  SWPIN,PTBD,WAITPRESS ;polling, wait for push
22   WAITRLSE:  sta   WATCHDOG        ;reset watchdog
23              brclr  SWPIN,PTBD,WAITRLSE ;polling, wait for release
24              rts
```

Code Listing 1.2: Driver for Normally-Open Push-Button Circuit on a PTB Pin.

The driver interface subroutine ISPUSHED uses a BRCLR instruction to implement an IF statement that sets the carry flag if the push-button is pushed (input pin is clear), otherwise clears it.

The driver interface subroutine WAITPRESS is an example of polled I/O using the value on the data register to implement the polling. The software implements the polling loop on lines 20 and 21. It resets the COP watchdog counter in the polling loop because there is no way to know when the button will be pressed. The subroutine subsequently enters a second polling loop until the switch is released. WAITPRESS is an example of a blocking subroutine; the subroutine will not return control back to the caller until a press-and-release event occurs on the switch.

1.3.4   SWITCH BOUNCE

Switch bounce is a phenomenon that occurs in switches due to characteristics of their mechanical construction. Switch bounce causes the switch to make and break contact several times before settling to its final state. Some factor that can lead to switch bounce are finite switch mass, excessive switch "springiness" and increased contact resistance cause by dirt or corrosion. Whatever the causes of switch bounce, the common result is that the digital output of the switch circuit alternates between 0 and 1 several times each time there is a switch event (opening or closing of the switch). In a digital circuit, this switching can occur over a time interval that is very long compared to the period of the CPU clock, often as long as several milliseconds. A program using the switch input may erroneously detect multiple transitions for each switch event.

Figure 1.11 illustrates the effect of switch bounce on a 0-to-1 switch transition. In the figure, the arrows mark the points at which the CPU is reading the switch input. The CPU should see the switch event as a series of 0's followed by a series of 1's, in this case $(0, 1, 1, 1)$, indicating the switch sequence *open-to-closed*. Because the CPU happens to read the switch on a bounce, the sequence of

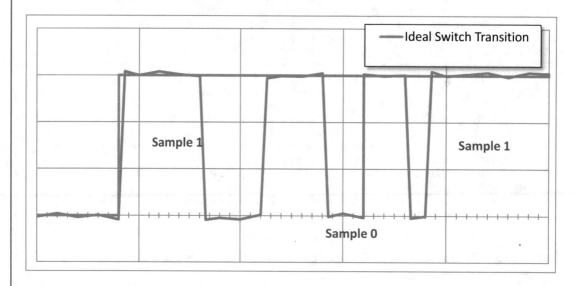

Figure 1.11: Oscilloscope view of switch bounce.

inputs read is $(0,1,0,1)$, which the software would interpret as the switch events *open-to-closed-to-open-to-closed*. Since the CPU has no way to distinguish a bounce form a real switch event, the program erroneously detects multiple events.

There are both hardware and software solutions for dealing with switch bounce. The most common hardware-based solution is to use a resistor-capacitor circuit in a low-pass filter configuration to smooth out the bounce transitions, as shown in Figure 1.12. The value of C must be chosen

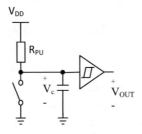

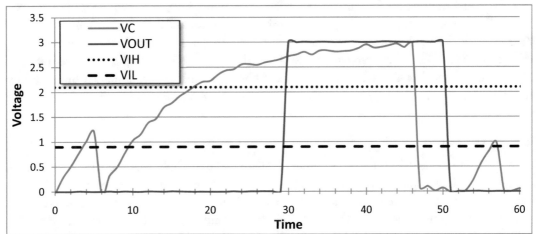

Figure 1.12: Hardware Debouncing Switch Interface Circuit.

such that the time constant $(R \times C)$ is long compared to maximum time between bounces, which prevents the input voltage from reaching the logic-high value unless the switch has been open for a "long time". When the switch closes, the capacitor quickly discharges through the low-impedance path to ground (a small resistor can be placed in series with the switch to limit this current). The Schmitt buffer, with its large input hysteresis, is needed to prevent the long rise time of the resistor-capacitor circuit from holding the input pin at non-logic levels during switching.

The simplest software solution to switch bounce is to ensure that the switch is read at most once during the switch bounce interval. This can be accomplished using delay synchronization by ensuring that the time between reads is greater than the bounce interval, T_{bounce}. This is done by inserting a delay between reads such that $T_{delay} > T_{bounce}$; this delay can be inserted anywhere between reads (that is, the driver can delay before each read or after each read). Although the program might still read the switch state during the bounce interval, it can read it at most once during each bounce interval and the value read must match the value of the input either before or after the switch event; that is, the program will always read the sequence $(0,0,1)$ or $(1,1,0)$ and correctly detect at most one switch transition. For this approach to work, it is necessary to find a delay that satisfies the condition $T_{delay} > T_{bounce}$, either through measurements of switch behavior or simple trial-and-error.

Code Listing 1.3 illustrates the use of delay synchronization in a switch driver to eliminate the effects of switch-bounce. The bounce interval is assumed to be less than 1 ms; thus, a delay of 1 ms is used. Note that the only change made to the previous switch driver routines is to call the delay subroutine each time the switch is read. A software delay loop, in subroutine DELAY1MS, is used to achieve the required 1 ms delay. Assuming the CPU bus clock is 4 MHz, a 1 ms delay is equivalent to 1 ms $\times$ 4 MHz= 4000 CPU clock cycles. A 16b counter loop using HX as the counter is used. Counting the CPU instruction cycles, this subroutine requires $24 + 8 \times$ ITERATIONS cycles to execute (including the BSR). Equating this to 4000 gives exactly ITERATIONS=497, where ITERATIONS is the number of loop iterations executed (the nop was added to the loop to tweak the timing to get ITERATIONS as an integer value). Thus, the execution of this subroutine, from BSR to RTS, is exactly 1 ms given a 4 MHz system clock. Although in this application exact timing is not necessary, being precise here allows this subroutine to be reused for other applications.

1.3.5   LED INDICATORS

A light-emitting diode (LED) is a semiconductor device that emits light when suitably driven by an external circuit. In general, a diode is a device that conducts current better in one direction (forward direction) than another (reverse direction); thus, an ideal diode acts like a one-way valve to current flow. When connected to an external circuit such that current flows in the forward direction, the diode is said to be forward-biased. When forward biased, the diode has an exponential current-voltage relationship; very little current flows in the forward direction until the external voltage applied across the diode exceeds a "turn-on" or threshold value, called the forward threshold voltage, V_f. A simplified model of a diode assumes that the diode does not conduct current (is "off") until the applied voltage is at least V_f; beyond that point, the diode approximates a short-circuit in the forward direction. Note that it is the external circuit that determines the operating region or bias of the diode. When operated in the forward-bias ("on") region, the external circuit must limit the current that flows through the diode; if the current exceeds the maximum rated value for the diode, the diode will "burn-out."

When operated in the "on" region, a LED emits light. The current that flows through the LED determines the luminous intensity. A simple LED circuit is shown in Figure 1.13. The series resistor is necessary only to limit the current that flows through the device; it can be connected between the anode and V_{DD} or between the cathode and ground. The value of the current-limiting

```
1    PTBD        equ   $02              ;port B data register
2    PTBDD       equ   $03              ;PTB direction control register
3    PTBPE       equ   $1844            ;PTB pull-up control register
4    SWPIN       equ   3                ;defines the pin that switch is on
5    ;------------------------------------------------------------------
6    INITSW:     psha                   ;callee save
7                bclr  SWPIN,PTBDD      ;only change setting for this pin
8                lda   PTBPE            ;enable internal pull-up resistor
9                ora   #(1<<SWPIN)      ;using an OR mask
10               sta   PTBPE            ;write back modified PTBPE value
11               pula                   ;callee restore
12               rts                    ;
13   ;------------------------------------------------------------------
14   ISPUSHED:   bsr   DELAY1MS         ;debounce
15               brclr SWPIN,PTBD,PRESSED ;detect if pressed
16               clc                    ;return C=0 if not
17               bra   END_ISP
18   PRESSED     sec                    ;else return C=1
19   END_ISP     rts
20   ;------------------------------------------------------------------
21   WAITPRESS:  sta   WATCHDOG         ;reset watchdog
22               bsr   DELAY1MS         ;debounce
23               brset SWPIN,PTBD,WAITPRESS ;polling, wait for push
24   WAITRLSE:   sta   WATCHDOG         ;reset watchdog
25               bsr   DELAY1MS         ;debounce
26               brclr SWPIN,PTBD,WAITRLSE ;polling, wait for release
27               rts
28   ;------------------------------------------------------------------
29   ITERATIONS  equ   497    ;# iterations to achieve 1ms delay for 4MHz bus
30   DELAY1MS    pshx           ;callee save
31               pshh
32               ldhx  #ITERATIONS ;load loop counter
33   DELAYLOOP   aix   #-1        ; decrement counter
34               cphx  #$0000     ; compare with 0
35               bne   DELAYLOOP  ; continue until 0
36               pulh             ; callee restore
37               pulx
38               nop              ;extra cycle to get exactly 1ms
39               rts
```

Code Listing 1.3: Software Debounced Normally-Open Push-Button Switch Driver.

resistor can be determined by setting $V_{diode} = V_f$ and computing R_{CL} such that I_{diode} does not exceed the maximum allowed for the device. Since, $V_{DD} - V_R = V_{diode}$ it follows that $V_{DD} - V_R = V_f$, or $V_R = V_{DD} - V_f$. Substituting $V_R = I_{diode}R_{CL}$ and rearranging, the diode current is related to R_{CL} (approximately) by $I_{diode} = (V_{DD} - V_f)/R_{CL}$. Given this, R_{CL} is chosen to ensure that I_{diode} is within the operating range for the diode.

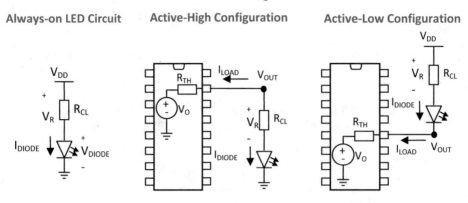

Figure 1.13: Always-on, active-high and active-low LED circuit configurations.

Example 1.4. Compute a value for the current limiting resistor in the always-on circuit in Figure 1.13, assuming $V_f = 1.6V$, $V_{DD} = 3.3V$ and a diode current of 5-10 mA is needed.
Solution: For $I_{DIODE} = 5$ mA, $R = (3.3 - 1.6)/(5$ mA$) = 340$ Ω.
For $I_{DIODE} = 10$ mA, $R = (3.3 - 1.6)/(10$ mA$) = 170$ Ω.
Answer: value of R must satisfy 170 Ω $< R < 340$ Ω.

To control a LED with a general-purpose output pin, the output is used to provide the V_{DD} (active-high) or ground (active-low) reference point in the circuit, as shown in Figure 1.13. Ideally, when the output pin is set to logic 0, the pin is at 0V; when the output is set to logic 1, the pin is at V_{DD} volts. In the active-low circuit, the LED is on when the output voltage is low and off when the output voltage is high. In the active-high configuration, the circuit has the opposite behavior (the LED is on when the output is high). When the LED is on, the diode current is flowing into the pin in the active-low configuration and out of the pin in the active-high configuration. The pin is said to "sink" the current in the active-low configuration and "source" it in the active-high configuration. The non-ideal behavior of V_{OH} and V_{OL} must be considered when selecting a value for the current-limiting resistor to obtain sufficient diode current to maintain brightness. Effectively, the value of R_{CL} must be reduced by the value R_{TH}.

Example 1.5. Compute a value for the current limiting resistor for the active-low configuration, assuming $V_f = 1.6V$ and a diode current of 10 mA is desired to get adequate brightness; assume $V_{DD} = 3.3V$. Assume the diode is connected to PTA1 with high drive strength.

Solution: Ignoring the effect of V_{OL}, For $I_{DIODE} = 10$ mA, $R = (V_{DD} - V_F)/I_{DIODE} = (3.3 - 1.6)/(10$ mA$) = 170$ Ohm. At high drive strength, the Thevenin equivalent resistance is approximately 25 Ω. Thus, $R = 170 - 25 = 145$ Ω.

Answer: value of R must be close to 145 Ohm.

Code Listing 1.4 shows an example of a driver for a single LED connected in the active-low configuration to pin PTB3. This driver code assumes that PTBD, PTBDD, PTBPE and PTBSE have

```
 1    ;--------------------------------------------------------------
 2    ;Driver for a active-low LED circuit interfaced on PTB3.
 3    LED_PIN     equ  3         ;pin 3
 4    PTBD        equ  $0002     ;port B data
 5    PTBDD       equ  $0003     ;port B DDR
 6    PTBSE       equ  $1845     ;port B slew control
 7    PTBDS       equ  $1846     ;port B drive strength
 8    ;-------------INIT_LED Initialization Subroutine----------------
 9    INIT_LED    psha
10                lda  PTBDS     ;high drive strength
11                ora  #(1<<LED_PIN)
12                sta  PTBDS
13                lda  PTBSE     ;no slew rate control
14                and  #~(1<<LED_PIN)
15                sta  PTBSE
16                bset LED_PIN,PTBDD ;configure GPIO for output
17                bsr  LED_OFF       ;call LED_OFF to turn off LED
18                pula
19                rts
20    ;------------LED_ON Subroutine-----------------------------------
21    LED_ON      bclr  LED_PIN,PTBD ;set pin to '0' to turn LED on
22                rts
23    ;------------LED_OFF Subroutine----------------------------------
24    LED_OFF     bset  LED_PIN,PTBD ;set pin to '1' to turn LED off
25                rts
26    ;------------TOGGLE_LED Subroutine------------------------------
27    TOGGLE_LED  psha                 ;callee save
28                lda  PTBD            ;get port data register
29                eor  #(1<<LED_PIN)   ;toggle bit corresponding to LED
30                sta  PTBD            ;write back to data register
31                pula
32                rts
```

Code Listing 1.4: Driver for Active Low LED Circuit.

been defined. Four driver interface routines are provided. INIT_LED configures the pin on PTB3 to be an output and calls LED_OFF interface routine to ensure the LED always starts in the off state. Even though only one instruction is required to turn off the LED, it is good practice to keep the code modular and call the LED_OFF subroutine to do this. In this way, if the hardware configuration is changed to active-high, only the drive code needs to be modified. LED_ON and LED_OFF are called to turn the led on or off, respectively. TOGGLE_LED toggles the state of the LED by loading the port data register into accumulator A, using a EOR-mask to toggle the bit corresponding to the output pin and storing the result back to the port data register. Instead of defining a separate constant to represent the mask, the assembler is called upon to generate the mask from the pin number using the shift left assembler operator; the expression $(1 << \text{LED\_PIN})$ is evaluated to a constant by the assembler and does not create any new instructions. In this instance, LED_PIN is 3, and 1 shifted left by 3 is the constant %00001000, which is the EOR mask to toggle bit 3.

A seven-segment display is an array of LEDs that share a common electrode (anode or cathode). Recall that the anode is the electrode into which current flows. Figure 1.14 shows the circuit configuration of a common-cathode seven-segment display module and its interface to Port B. Note that each LED segment requires it own current-limiting resistor (a single shared resistor connected in series with the common terminal will cause variable brightness on the displays and possible lead to excess current if a single LED is on). In addition to computing the correct resistor values, the maximum combined I_{OLT} (60 mA) must not be exceeded when all LEDs are on; assuming no other loads are being driven low by the microcontroller, the maximum individual LED current must not exceed $|I_{OLT}|/7$, or approximately 8.5 mA per segment.

The driver for the seven segment display is shown in Code Listing 1.5. This driver corresponds to the seven segment circuit interfaced as shown in Figure 1.14, connected on PTB6–PTB0 in an active-high configuration. The entire driver is designed not to affect pin PTB7, allowing another driver to control pin PTB7. The INIT_SSEG initialization subroutine configures pins PTB6 through PTB0 as output without affecting the state of pin PTB7. It does this using an OR-mask to set the PTBDD bits used by the interface. The initialization subroutine also calls the driver subroutine SSEG_OFF to initialize the seven segment state to all off, so that it starts in a blank state. Driver subroutine SSEG_OFF uses an AND-mask to clear the bits of the Port B data register corresponding to the interface.

Driver subroutine OUT_DIG uses a lookup-table to output the correct seven-segment display code based on the decimal digit passed in accumulator A. After the program checks that the input parameter represents a valid digit (Lines 35-36), the parameter is transferred to HX. This allows the lookup table to be indexed by the digit parameter to retrieve the correct code. The seven code bits retrieved from the table need to be placed in the least significant 7 bits of the Port B data register without affecting the most significant bit. This is accomplished by loading the current value from the Port B data register (Line 39), clearing the least significant 7 bits with an AND-mask (Line 40), and combining this with the code data using an ORA instruction (Line 41). The modified Port B data register value is then written back to the interface register (Line 42). The remainder of the subroutine computes the return value, which sets the carry to indicate an error (invalid BCD digit) or clear it to indicate no error.

```
1    SSEG_1MSK   equ   %01111111  ;SSEG pins 1 mask
2    SSEG_0MSK   equ   %10000000  ;SSEG pins 0 mask
3    PTBD        equ   $02        ;port B data
4    PTBDD       equ   $03        ;port B DDR
5    PTBSE       equ   $1845      ;port B slew rate enable
6    PTBDS       equ   $1846      ;port b drive strength
7    ;--------------------INIT_SSEG Subroutine------------------------
8    ; Configures pins PTB6-0, Turns off all segments
9    INIT_SSEG   psha             ;callee save
10               bsr   SSEG_OFF   ;set initial output to all off
11               lda   PTBSE      ;get current SE config
12               and   #SSEG_0MSK ;configure pins for no slew control
13               sta   PTBSE      ;write back modified DDR
14               lda   PTBDS      ;get current DDR config
15               ora   #SSEG_1MSK ;configure high drive strength
16               sta   PTBDS      ;write back modified DDR
17               lda   PTBDD      ;get current DDR config
18               ora   #SSEG_1MSK ;configure SSEG pins as output
19               sta   PTBDD      ;write back modified DDR
20               pula             ;callee restore
21               rts              ;return
22   ;--------------------SSEG_OFF Subroutine------------------------
23   ;Sets PTA6-PTA0 to 0 to turn off segments
24   SSEG_OFF    psha             ;callee save
25               lda   PTBD       ;get current port data
26               and   #SSEG_0MSK ;turn off seven segment bits
27               sta   PTBD       ;write modified port data
28               pula             ;callee restore
29               rts              ;return
30   ;--------------------OUT_DIG Subroutine------------------------
31   ;Outputs seven segment code of the digit passed in accumulator A
32   OUT_DIG     psha             ;callee save
33               pshx
34               pshh
35               cmp   #$0A       ;check digit
36               bhs   SSEG_ERR   ; if >9 goto error
37               clrh
38               tax              ;HX=digit
39               lda   PTBD       ;load port data
40               and   #SSEG_0MSK ;zero out bits corresponding to SSEG
41               ora   SSEG_TAB,X ;combine SSEG data from lookup table
42               sta   PTBD       ;write modified port data
43               clc              ;clear error flag
44               bra   RTN        ;goto return
```

Code Listing 1.5: Common-Cathode Seven Segment Display Driver for Figure 1.14 (*Continues*).

```
45   SSEG_ERR    sec                  ;set error from earlier
46   RTN         pulh                 ;callee restore
47               pulx
48               pula
49               rts                  ;return
50   ;---------------Seven Segment Code Constant Array------------------
51   SSEG_TAB    dc.b   $7E,$30,$6D,$79,$33,$5B,$5F,$70,$7F,$73
```

Code Listing 1.5: (*Continued*) Common-Cathode Seven Segment Display Driver for Figure 1.14.

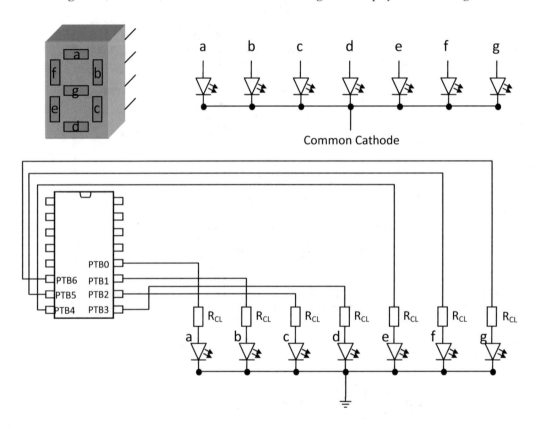

Figure 1.14: Common-cathode seven-segment display interface.

1.3.6   EMULATION OF OPEN-DRAIN AND TRI-STATE OUTPUTS

Open-drain and tri-state logic devices include a high-impedance output state, usually denoted by 'Z'. Open-drain devices have two output states, '0' and 'Z', and cannot drive an output to a logic '1' directly (a pull-up resistor is required). Open-drain is the term used when the integrated circuit technology is based on field-effect transistors, while open-collector is used when the technology is based on bipolar junction transistors. Open-drain outputs are often used for creating wired-AND and wired-OR logic or for interfacing logic at different voltage levels. A tri-state logic element can output all three states: '0', '1' and 'Z'.

A general-purpose I/O pin can emulate the high-impedance state by configuring the pin as an input. To the external circuit, the input pin presents high impedance (almost open circuit). To drive the output high or low, the pin is configured as an output driving high or low; to drive the output to 'Z', the pin is configured for input.

Code Listing 1.6 illustrates a simple driver to emulate an open-drain output on PTB3. The

```
1    ;------------------------------------------------------------------
2    ; Driver to emulate Open Drain Output on PTB3
3    OC_PIN    equ   3     ;pin 3
4    PTBD      equ   $02   ;port B data
5    PTBDD     equ   $03   ;port B DDR
6    ;-----------------INIT_OC Subroutine-------------------------
7    ; Configures pins PTB3 as Hi-Z
8    INIT_OC   bsr   OUT_Z;set initial value to output 'Z'
9              rts
10   ;-------------------OUT_X Subroutine-------------------------
11   ; Configures PTB3 as input to make it 'output' high impedance
12   OUT_Z     bclr  OC_PIN,PTBDD   ;configure GPIO to output 'Z'
13             rts
14   ;-------------------OUT_0 Subroutine-------------------------
15   ; Configures pins PTB3 output driving low
16   OUT_0     bclr  OC_PIN,PTB    ;set pin output to '0'
17             bset  OC_PIN,PTBDD  ;then program pin as output
18             rts
```

Code Listing 1.6: Driver to Emulate Open Drain Outputs on GPIO Pins.

initialization subroutine, starting on line 8, calls the driver subroutine OUT_Z to output a high-impedance on the pin before returning. The subroutine OUT_Z configures the pin as input to emulate the high-impedance output state. The subroutine OUT_0 clears bit 3 of the Port B data register before configuring the pin as output, to prevent a possible glitch on the output. If the pin is configured as output first, an erroneous value in Port B data register could cause a temporary short-circuit condition if another external source is driving the output low (in a wired-or configuration, for example).

Open-drain outputs may require use of an external pull-up resistor to achieve the logic high state. The pull-up resistor should be chosen to be large enough to reduce the effect that I_{LOAD} has on V_{OL} when the pin is configured to output a logic '0'; at the same time, the resistor must not be large enough to have a significant voltage drop caused by the I_{IL} leakage current when the pin is configured as input. Tri-state outputs do not require the external resistor as the output can drive both high and low logic levels (as well as the high impedance). If multiple tri-state outputs are connected to a common circuit node, some coordination is required to ensure that only one is driving the node at any given time. Tri-state driver code is left as an exercise.

1.4   INTERRUPT SYNCHRONIZATION

Interrupt synchronization uses a CPU hardware mechanism to interrupt normal instruction sequencing so that an I/O event can be serviced. When an I/O device or peripheral needs the CPU to perform an input or output operation, it triggers this hardware mechanism via the I/O port interface to issue an *interrupt request* to the CPU. The CPU can be configured to temporarily stop executing the current instruction sequence and run a special interrupt service routine (ISR) to service the I/O request. Upon completion of the ISR, the CPU resumes execution of the interrupted instruction sequence. Interrupt synchronization is useful when I/O events happen sporadically since the CPU does not waste excessive processing cycles polling.

1.4.1   HCS08 CPU INTERRUPT PROCESSING

Details for interrupt handling on specific I/O port interfaces can vary significantly. However, since the HCS08 CPU processes all interrupt requests, the basic interrupt processing sequence is independent of interface details. This common processing includes how interrupt service routines are located, what happens when multiple interrupt requests are made, and how the CPU restores registers contents and returns to the interrupted instruction sequence.

Each port interface that can issue interrupt requests to the HCS08 CPU is associated with a vector in the interrupt vector table. This vector identifies the start of the service routine for the interrupt. When an interrupt request occurs, the CPU fetches the address of the associated service routine from the vector table and places this value in the PC, which effectively implements a jump to the ISR. After the ISR completes, the CPU needs to resume execution of the interrupted instruction sequence. This requires that the return address be stacked before the vector is fetched from the vector table. Furthermore, for the interrupted sequence to resume correctly, the values of the CPU registers must be restored. Thus, before executing the ISR, the CPU also stacks the registers (automatic callee-saving). A special return from interrupt instruction, RTI, unstacks the registers and the PC, which returns the CPU registers to their exact state before the CPU was interrupted.

It is possible for multiple interrupt requests to occur simultaneously. The interrupt sources have been prioritized to resolve which interrupt is serviced first. The interrupt vectors have been assigned such that the higher the vector location in memory, the higher the interrupt priority. For example, Table 1.1 shows that the analog to digital converter interrupt vector is located at $FFD8 while the interrupt request pin vector is at $FFFA; thus, the interrupt request pin has a higher priority than the analog to digital converter. It is also possible for an interrupt request to occur during execution of an interrupt service routine; this is called nested interrupts. To prevent nested interrupts, the I mask bit in the CCR, when set, prevents subsequent interrupt requests from being passed on to the CPU. Masked requests are queued by the CPU and serviced in priority order upon clearing of the interrupt mask.

The HCS08 CPU interrupt sequence consists of the following steps. First, CPU registers are stacked. The stacking order is PC first, followed by X, A and CCR. Note that H is not stacked by the CPU. If the ISR modifies H, it must be explicitly callee saved in the ISR. Second, the I mask bit in the CCR becomes set to prevent nested interrupts. Third, the CPU fetches the vector of the highest priority pending interrupt request into the PC, jumping to the ISR.

At the end of the ISR, the RTI instruction is used to unstack CCR, A, X and PC. Unstacking the PC effects the return to the interrupted instruction sequence. Unstacking CCR restores the I mask in the CCR to 0, enabling subsequent interrupt requests.

In addition to the global interrupt mask bit in the CCR, each interrupt source has a local enable bit in one of its configuration registers. The driver using the interface has the responsibility of enabling interrupts for the interface if interrupt synchronization will be used. These local masks mean that interrupt service routines do not need to be created for unused interfaces or those that will use polling or delay synchronization.

It is usually necessary for the interrupt service routine to acknowledge the interrupt request in order to clear it. If the request is not acknowledged, it remains asserted upon completion of the ISR, which immediately causes the request to be repeated. Clearing the request may be an explicit action, such as writing to a specific bit in a control register, or be performed implicitly as a side-effect of accessing I/O ports.

Data exchange between the main program and an ISR is only possible through shared global variables (the stack and CPU registers cannot be used because it is not known exactly when the ISR will execute). Care must be taken when the main program is accessing a shared global variable since the interrupt can occur at any time. For example, if the main program loads from a shared global variable and performs a computation with the value, another interrupt can occur before the modified value is written back to the variable location. The main program would then be working with a stale value or, even worse, the newer value from the ISR could be overwritten causing an input to be lost. To prevent such coherence problems, it is best to mask interrupts before accessing a global variable that is shared with an ISR and then unmask them when all computations with the value are complete. Since interrupts are masked when the ISR is running, there is no such problem accessing the value from within the ISR.

1.4.2 IRQ INTERRUPT PIN

The pin PTA5 can be configured as an active low external interrupt request (IRQ) pin with an internal pull-up resistor. When configured in this manner, an external peripheral can be integrated into the microcontroller system using interrupt synchronization. In order to enable the IRQ function on this pin, the higher priority functions associated with this pin (active low CPU reset input and external timer module clock) must be disabled. If a mechanical switch is to be integrated to trigger an IRQ, the switch should be hardware debounced; the long delays due to software debouncing in the ISR will not only delay servicing other interrupts, but may cause the watchdog timer to expire. It is usually not a good idea to reset the watchdog counter within an ISR because even if the main program enters an infinite loop, interrupts can still occur. In this case, the main loop remains in the infinite loop while the ISR feeds the watchdog.

The IRQ pin can be configured to be falling-edge triggered or low-level triggered. When configured as falling-edge triggered, an interrupt request is only made on a high-to-low logic level transition on the pin; the external circuit must return the pin to the unasserted (high) state before making another interrupt request. When configured as level-triggered, interrupt requests are continually made as long as the IRQ pin is asserted (held low); the external circuit does not have to return the pin to the deasserted level to issue another request.

Configuration of the IRQ pin is performed via the Interrupt Pin Request Status and Control register (IRQSC). The format of this register is shown in Figure 1.15. IRQPDD (IRQ Pull Device

IRQSC: Interrupt Request Pin Status and Control 2 (memory-mapped at address $000F)

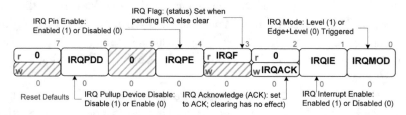

Figure 1.15: Interrupt Request Status and Control Register.

Disable) can be set to disable the internal pull-up resistor on the IRQ pin. IRQPDD is 0 by default, meaning the internal pull-up is enabled. IRQPE (IRQ Pin Enable) must be set to enable the IRQ function on pin 1 (disabled by default). IRQACK (IRQ ACKnowledge) is a write-only bit that is used to acknowledge the IRQ interrupt request, indicating that the external device has been serviced. Reading IRQACK always returns 0. The ISR for the IRQ pin must write a 1 to IRQACK (using a BSET instruction, for example) to clear the interrupt request. The IRQIE (IRQ Interrupt Enable) is the local enable for the IRQ interrupt request. This bit must be set to have an interrupt request forwarded to the CPU; when clear, the IRQ pin can still be used as an active low input by polling IRQF, but no interrupt will ever be triggered. Finally, IRQMOD (IRQ MODe) selects whether the IRQ pin will be falling-edge triggered (IRQM=0) or low-level triggered (IRQM=1).

Writing configuration data to the IRQ configuration register can change pin states due to enabling pull-up devices. These state changes can trigger a false interrupt request. Thus, modifications to the IRQ control bits should be done with the global interrupt mask set to prevent a false interrupt from causing the service routine to execute before the IRQ driver is finished initializing. In addition, it is a good idea to clear any pending false interrupt requests by writing a 1 to the IRQACK bit after all changes have been made to IRQ control registers.

Code Listing 1.7 demonstrates a skeleton MC9S08QG4/8 interrupt-based driver for the IRQ pin. The driver has been designed to perform a simple data exchange with the main program. Each time an interrupt request is made, the ISR increments a global variable called IRQCOUNT. The main program loop repeatedly tests this variable and decrements it if it is nonzero. Though simple,

```
1   IRQSC         equ  $000F
2   ;---------Required Global Variables (to be defined in ram)-------
3   ;IRQCOUNT   ds.b  1           ;global variable for data exchange w/ISR
4   ;------Driver Initialization------------------------------------------
5   IRQINIT       clr  IRQCOUNT   ;initialize driver variable
6                 mov  #$12,IRQSC ;IRQ pin enabled; edge triggered
7                                 ;interrupts and pull-ups enabled
8                 bset 2,IRQSC    ;clear any pending interrupts
9                 rts
10  ;------ISR for IRQ----------------------------------------------------
11  IRQ_ISR       inc  IRQCOUNT   ;increment global variable
12                bset 2,IRQSC    ;clear any pending interrupts
13                rti             ;return from ISR
14  ;--------------------------------------------------------------------
```

Code Listing 1.7: Simple Interrupt-Based Driver for IRQ Pin.

this example illustrates most of the basic principles of interrupt processing. Line 3 defines the global variable that will be used for data exchange between the ISR and the main program. This variable is initialized in the driver initialization routine IRQINIT. Line 5 marks the start of the initialization subroutine. The shared global variable is first initialized to 0. Then, the IRQSC register is written to configure the IRQ pin. On line 8, the IRQACK bit in IRQSC is set to acknowledge any pending interrupt request that may have been triggered while configuring the IRQ pin.

    The interrupt service routine begins on line 11. The main function of the ISR, incrementing the global counter variable IRQCOUNT, is done first. Then, the interrupt request is cleared (acknowledged) by writing a 1 to the IRQACK bit in IRQSC. Finally, an RTI instruction returns control to the interrupted program. Example 1.6 illustrates how the shared global variable is accessed from software.

Example 1.6.   Write the instruction sequence that performs a saturating decrement on the global variable IRQCOUNT shown in Code Listing 1.7.
Solution: Before the program accesses the variable, interrupts must be masked to ensure the access is atomic. A saturating decrement never allows the result to go below zero; therefore the variable is tested for zero and is decremented if nonzero (lines 3 to 5). Interrupts are re-enabled after the sequence (line 6).
Answer:

```
1              sei             ;mask interrupts for atomic access
2              lda   IRQCOUNT  ;load shared variable
3              beq   IS_ZERO   ;if zero, do nothing
4              deca            ;else decrement
5              sta   IRQCOUNT  ;and update variable
6  IS_ZERO: cli               ;re-enable interrupts
```

When the IRQ pin function is enabled but IRQ interrupts are disabled (IRQIE=0), the IRQ pin can be used as an asynchronous active low input pin. The IRQF (IRQ Flag) bit in IRQSC, a read only status bit (writing to this bit has no effect), becomes set when an active low signal has been detected on the IRQ pin. This flag can be tested to determine if an IRQ event has occurred, even if the event happened in the past. In other words, the CPU does not have to be polling the IRQ pin at the time the event occurs. Thus, the IRQ pin can be used without interrupt synchronization to detect active low inputs. If multiple IRQ events occur before the CPU reads the IRQF flag, though, the CPU will only detect one event. Unlike using interrupt synchronization, there is no way to count how many active low triggers have occurred.

Code Listing 1.8 illustrates a non-interrupt based driver for the IRQ pin. Line 5 begins the

```
IRQSC         equ    $000F
;*********************************************************************
; Interrupt-based IRQ pin driver                                    *
;*********************************************************************
IRQINIT:      mov  #$10,IRQSC    ;IRQ pin enabled; edge triggered
                                 ;no interrupts and pull-ups enabled
              rts
IRQ_CHECK:    brclr 3,IRQSC,NO_IRQ ;test IRQF
              bset  2,IRQSC       ;acknowledge and clear IRQF if set
              sec                 ;set return value true
              bra   RETURN
NO_IRQ:       clc                 ;return value false
RETURN:       rts                 ;return
```

Code Listing 1.8: Non-Interrupt Based Driver for IRQ Pin.

code for the initialization subroutine, IRQINIT. This subroutine simply configures the IRQSC for IRQ pin enabled with interrupts disabled before returning. IRQ check is a subroutine that tests the IRQF flag and returns with the carry flag set if IRQF is 1 and clear otherwise. To clear the IRQF so that subsequent events can be detected, a 1 is written to the IRQACK bit in IRQSC.

1.4.3   KEYBOARD INTERRUPT

The keyboard interrupt (KBI) module allows up to 8 external signals to generate interrupt requests but share a single interrupt vector. The module is so named because it was originally intended to interface keypads and other switch arrays into a microcontroller system. Although the 8 external triggers share a common interrupt vector, the interrupt requests can be individually configured as either active high or low sources and collectively configured as either edge or level triggered. As a result, the pins can be used to allow up to 8 independent interrupt sources to be interfaced as long as the common ISR is designed to prioritize and service each source as appropriate. Such a service routine can sense the individual KBI pins by reading the Port A and B data registers.

The KBI module shares pins with port A and port B. KBI0-KBI3 share pins with Port A pins 0-3, while KBI pins 4-7 share pins with Port B pins 0-3. The KBI functionality has priority over the GPIO functions, but to use the KBI function on any of these pins the higher priority functions must be disabled.

The KBI status and control registers are shown in Figure 1.16. KBISC is the primary status

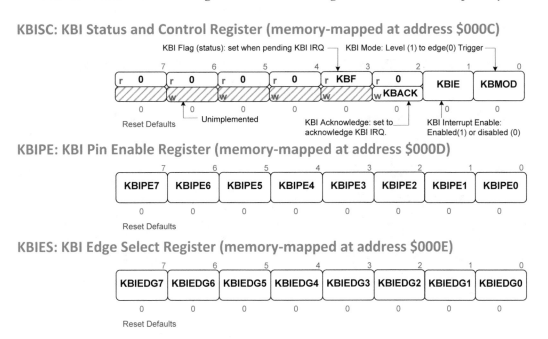

Figure 1.16: Format of the Keyboard Interrupt Status and Control Registers.

and control register for the KBI module. The control bits and status flag in this register are similar to the control and status pins for the IRQ module. KBF is the status flag that indicates that a KBI event has been detected on a pin. If KBIE (interrupt enable) is set, an interrupt request is generated in response to the event; otherwise, no interrupt request is generated and a non-interrupt based driver can test KBF. Writing a 1 to KBACK acknowledges the interrupt and clears the KBF flag. Note that, like the IRQ pin, it is not possible to determine if multiple KBI events have occurred since the flag was last cleared. KBMOD selects whether the KBI pins will be edge or level sensitive; edges are detected when KBMOD is clear.

The KBI Pin Enable (KBIPE) register contains eight individual enable bits that are used to individually select which KBI pins are active. Setting a bit of this register enables the corresponding KBI pin. KBI Edge Select (KBIES) allows each pin to be configured as active high or low (setting a bit selects active high or positive edge triggered mode, depending on KBMOD is KBISC). Enabling the pull-up resistor on the corresponding GPIO pin (in PTAPE or PTBPE registers) enables the internal register for the KBI pin. Unlike the GPIO function, however, the KBI can be a pull-up or pull-down resistor, depending on the polarity of the bit in the KBIES register. Thus, if PTAPE bit 0 is set (port a pin 0 pull-up enabled) and the polarity of KBI pin 0 is configured as active high, an internal pull-down resistor will be enabled instead of a pull-up resistor.

Like the IRQ pin, interrupts should be globally masked before configuring the KBI module to prevent a false interrupt from being triggered during configuration. After all control registers have been configured, the KBACK bit in KBISC should be set with a BSET instruction to clear any pending false interrupt requests.

Figure 1.17 shows the configuration of a keypad switch and its integration into a MC9S08QG8 using the KBI module. Internally, the keypad switch is typically configured as a

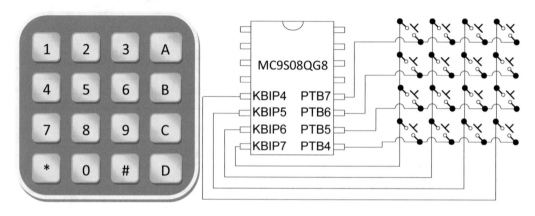

Figure 1.17: Configuration of a Keypad Switch and Its Integration to KBI Module.

matrix of normally-open push-button type switches. Each switch is connected to one row line and one column line such that there is one switch at the intersection of each row and column. In the configuration shown in Figure 1.17, each row is connected to a KBI input and each column is con- nected to a Port B output. Assuming that the KBI pull-ups are enabled, if no switch is pressed each row line will be high (due to the internal pull-ups) and no KBI interrupt is detected. If the Port B outputs are low, when one or more switches are pressed the low output is transmitted via the closed switch to one of the KBI inputs, triggering an interrupt.

To determine which key is pressed, the interrupt service routine can drive a logic low on the column lines, one by one, until one of the KBI lines (PTB4–PTB7) reads low. Assuming only one switch is pressed, its corresponding row input would be clear. The key number (0-15) can be computed as (4×row)+column. If two switches are pressed in a single row, then their column lines are electrically connected. Since one column line is driven low and one driven high, a short circuit will exist between the high and low pins which will cause excessive current to flow and may damage the microcontroller. To prevent this, the columns not being driven low should be driven to high impedance instead of high. If keys in multiple rows are pressed, this algorithm returns the first key found. It is possible to use a more complex algorithm to detect the individual keys of a multi-key press.

Code Listing 1.9 shows a driver for a keypad interfaced as shown in Figure 1.17. When multiple keys are pressed, priority is given to the key with the lowest column first, then the lowest row. The driver returns the ASCII value of the key in the global variable KEYVALUE. So that the main program can tell when there is a new key value, a Boolean variable ISKEY is set by the ISR when a new key value is place in KEYVALUE. The main program must clear ISKEY after reading the key value.

The driver initialization routine KBIINIT configures pins KBI4 to KBI7 as falling edge triggered inputs with KBI interrupt requests enabled. The internal pull-ups are also enabled. The initialization then configures PTB4 though PTB7 as outputs driving low; these low outputs allow the keypad to initially pass active low inputs to the KBI module to trigger an interrupt.

Driver subroutine DRCOL is a helper subroutine designed to drive a single column output (PTB4 through PTB7) at a low logic level, while the other column outputs are set to high impedance (configured as inputs). The subroutine uses a case structure to output the correct configuration data to the Port B data direction register based on the switch column parameter passed in index register X, which is the only parameter passed to the subroutine.

The interrupt service routine starts by looping through the columns, outputting low on each column one by one. On each iteration, the row data is read from Port B data and checked to see if any row lines read low. If so, the ISR continues at label FOUNDKEY. It is possible that no row lines are detected low for any column. This can happen due to switch bounce or if the CPU is prevented from servicing the KBI interrupt before the key has been released. If no key press is detected, the ISR acknowledges the interrupt and returns. The code at the label NOKEY can be modified to change this behavior.

Starting at the label FOUNDKEY, the code uses the row and column values to determine an ASCII value for the key that was pressed. At this point in the execution of the ISR, the column number is in index register X. The row code, which is in the lower nibble of A, is a value that either has a single zero in the bit whose number is the same as the row number, or multiple zeros if keys in multiple rows are pressed. Since there are only 15 values the row can have at this point (from 0000_2 to 1110_2) a lookup table is used to simultaneously prioritize the rows when multiple zero's are detected as well as compute $4 \times \text{row}$. In other words, the lookup table returns 0 when the row lines have value XXX0 (X is don't care); 4 when the row lines have value XX01; 8 when the row lines have value X011; and 12 when they have value 0111.

Once the key number has been computed as $4 \times \text{row} + \text{column}$, a lookup table is used to convert the key number into an ASCII value. This table can be modified to return any ASCII values desired, depending on what is shown on the keypad. Finally, the ASCII value is stored to KEYVALUE, ISKEY is set to true and the KBI interrupt is acknowledged.

1.5 CHAPTER PROBLEMS

1. List the three types of I/O control registers and briefly describe the purpose of each?

```
1    ;*********************************************************************
2    ; Interrupt Based KBI driver                                        *
3    ;*********************************************************************
4    KBIINIT      psha
5                 clr ISKEY            ;1 means unread key is in KEYVALUE
6                 mov #$02,KBISC       ;Enable KBI interrupt, edge triggered
7                 mov #$F0,KBIPE       ;enable KBIP4-7 pins
8                 mov #$00,KBIES       ;configure falling edge trigger
9                 lda #$0F
10                sta PTBPE            ;enable pull-up on PTB0-3 (KBIP4-7)
11                mov #$F0,PTBDD       ;config PTB4-7 as output(PTB3-0 are KBI)
12                clr PTBD             ;PTB data always 0 for this driver
13                bset 2,KBISC         ;clear any false KBI interrupt request
14                pula
15                rts                  ;return to caller
16   ;*********************************************************************
17   ;*DRIVECOL sets PTB4-7 to ZZZ0 (when X=0),ZZ0Z (when X=1),
18   ;    Z0ZZ (when X=2),0ZZZ (when X=3)
19   ;*********************************************************************
20   DRIVECOL:    cbeqx #3,DRCOL3      ;switch(X) case 3:
21                cbeqx #2,DRCOL2      ;case 2:
22                cbeqx #1,DRCOL1      ;case 1:
23   DRCOL0:      mov   #$10,PTBDD     ;default: output ZZZ0 to column lines
24                bra   ENDDRCASE
25   DRCOL1:      mov   #$20,PTBDD     ;output ZZ0Z to column lines
26                bra   ENDDRCASE
27   DRCOL2:      mov   #$40,PTBDD     ;output Z0ZZ to column lines
28                bra   ENDDRCASE
29   DRCOL3:      mov   #$80,PTBDD     ;output 0ZZZ to column lines
30   ENDDRCASE    rts
31   ;*********************************************************************
32   ;*ISR for KBI                                                       *
33   ;*********************************************************************
34   KBI_ISR      pshh                 ;H only register not auto stacked
35                clrx                 ;init column
36   ROWLOOP:     bsr  DRIVECOL        ;assert column i low, others hi-Z
37                lda  PTBD            ;load port B (row and column)
38                and  #$0F            ;mask off column
39                cmp  #$0F            ;check if any row lines low
40                bne  FOUNDKEY        ;goto found key if not all high
41                incx                 ;increment column counter
42                cpx  #$4
43                blo  ROWLOOP         ;repeat row loop until 0 shifts out
44                bra  NOKEY           ;handle when no key press detected
```

Code Listing 1.9: Interrupt Based Driver for a Keypad Interfaced to KBI Module *Continues).*

Code Listing 1.9 shows a driver for a keypad interfaced as shown in Figure 1.17. When multiple keys are pressed, priority is given to the key with the lowest column first, then the lowest row. The driver returns the ASCII value of the key in the global variable KEYVALUE. So that the main program can tell when there is a new key value, a Boolean variable ISKEY is set by the ISR when a new key value is place in KEYVALUE. The main program must clear ISKEY after reading the key value.

The driver initialization routine KBIINIT configures pins KBI4 to KBI7 as falling edge triggered inputs with KBI interrupt requests enabled. The internal pull-ups are also enabled. The initialization then configures PTB4 though PTB7 as outputs driving low; these low outputs allow the keypad to initially pass active low inputs to the KBI module to trigger an interrupt.

Driver subroutine DRCOL is a helper subroutine designed to drive a single column output (PTB4 through PTB7) at a low logic level, while the other column outputs are set to high impedance (configured as inputs). The subroutine uses a case structure to output the correct configuration data to the Port B data direction register based on the switch column parameter passed in index register X, which is the only parameter passed to the subroutine.

The interrupt service routine starts by looping through the columns, outputting low on each column one by one. On each iteration, the row data is read from Port B data and checked to see if any row lines read low. If so, the ISR continues at label FOUNDKEY. It is possible that no row lines are detected low for any column. This can happen due to switch bounce or if the CPU is prevented from servicing the KBI interrupt before the key has been released. If no key press is detected, the ISR acknowledges the interrupt and returns. The code at the label NOKEY can be modified to change this behavior.

Starting at the label FOUNDKEY, the code uses the row and column values to determine an ASCII value for the key that was pressed. At this point in the execution of the ISR, the column number is in index register X. The row code, which is in the lower nibble of A, is a value that either has a single zero in the bit whose number is the same as the row number, or multiple zeros if keys in multiple rows are pressed. Since there are only 15 values the row can have at this point (from 0000_2 to 1110_2) a lookup table is used to simultaneously prioritize the rows when multiple zero's are detected as well as compute $4 \times$ row. In other words, the lookup table returns 0 when the row lines have value XXX0 (X is don't care); 4 when the row lines have value XX01; 8 when the row lines have value X011; and 12 when they have value 0111.

Once the key number has been computed as $4 \times$ row+column, a lookup table is used to convert the key number into an ASCII value. This table can be modified to return any ASCII values desired, depending on what is shown on the keypad. Finally, the ASCII value is stored to KEYVALUE, ISKEY is set to true and the KBI interrupt is acknowledged.

1.5   CHAPTER PROBLEMS

1. List the three types of I/O control registers and briefly describe the purpose of each?

```
;*****************************************************************
; Interrupt Based KBI driver                                    *
;*****************************************************************
KBIINIT        psha
               clr ISKEY              ;1 means unread key is in KEYVALUE
               mov #$02,KBISC         ;Enable KBI interrupt, edge triggered
               mov #$F0,KBIPE         ;enable KBIP4-7 pins
               mov #$00,KBIES         ;configure falling edge trigger
               lda #$0F
               sta PTBPE              ;enable pull-up on PTB0-3 (KBIP4-7)
               mov #$F0,PTBDD         ;config PTB4-7 as output(PTB3-0 are KBI)
               clr PTBD               ;PTB data always 0 for this driver
               bset 2,KBISC           ;clear any false KBI interrupt request
               pula
               rts                    ;return to caller
;*****************************************************************
;*DRIVECOL sets PTB4-7 to ZZZ0 (when X=0),ZZ0Z (when X=1),
;    Z0ZZ (when X=2),0ZZZ (when X=3)
;*****************************************************************
DRIVECOL:      cbeqx #3,DRCOL3        ;switch(X) case 3:
               cbeqx #2,DRCOL2        ;case 2:
               cbeqx #1,DRCOL1        ;case 1:
DRCOL0:        mov   #$10,PTBDD       ;default: output ZZZ0 to column lines
               bra   ENDDRCASE
DRCOL1:        mov   #$20,PTBDD       ;output ZZ0Z to column lines
               bra   ENDDRCASE
DRCOL2:        mov   #$40,PTBDD       ;output Z0ZZ to column lines
               bra   ENDDRCASE
DRCOL3:        mov   #$80,PTBDD       ;output 0ZZZ to column lines
ENDDRCASE      rts
;*****************************************************************
;*ISR for KBI                                                   *
;*****************************************************************
KBI_ISR        pshh                   ;H only register not auto stacked
               clrx                   ;init column
ROWLOOP:       bsr DRIVECOL           ;assert column i low, others hi-Z
               lda PTBD               ;load port B (row and column)
               and #$0F               ;mask off column
               cmp #$0F               ;check if any row lines low
               bne FOUNDKEY           ;goto found key if not all high
               incx                   ;increment column counter
               cpx #$4
               blo ROWLOOP            ;repeat row loop until 0 shifts out
               bra NOKEY              ;handle when no key press detected
```

Code Listing 1.9: Interrupt Based Driver for a Keypad Interfaced to KBI Module *Continues).*

```
45  FOUNDKEY:    psha                   ;save row
46               clrh                   ;HX will index array, H must be $00
47               txa                    ;copy column number into A
48               pulx                   ;get row code in X
49               add   ROWTBL,X         ;convert to 4*ROW, add to column
50  TOASCII      tax                    ;copy key value (0-15) into X
51               lda   KEYTBL,X         ;convert key number to ASCII key value
52               bra   STORKEY
53  NOKEY        bra   ACK_KBI          ;just return if no key detected
54  STOREKEY     sta   KEYVALUE         ;store key value
55               bset  1,ISKEY          ;set ISKEY to true to indicate valid key
56               mov   #$F0,PTBDD       ;output 0 on all col lines for next key
57  ACK_KBI      bset  2,KBISC          ;acknowledge CPU interrupt request
58               pulh                   ;unstack h
59               rti                    ;return from ISR
60  ;ROW conversion table
61  ROWTBL       dc.b  $00,$04,$00,$08,$00,$04,$00,$0C
62               dc.b  $00,$04,$00,$08,$00,$04,$00
63  KEYTBL       dc.b  '1','2','3','A','4','5','6','B'
64               dc.b  '7','8','9','C','*','0','#','D'
```

Code Listing 1.9: (*Continued*) Interrupt Based Driver for a Keypad Interfaced to KBI Module.

2. Suppose a temperature sensor were to be connected to a microcomputer system. Of the three types of I/O interfaces, which type would be accessed in order to

   (a) Set up the sensor parameters, such as selecting units (Celsius of Fahrenheit)?

   (b) Read temperature value?

   (c) Determine if there are any sensor faults or other exceptional conditions?

3. Describe the difference between memory mapped I/O differ and separate I/O? List the main advantage of each type.

4. What is the purpose of the COP watchdog? What happens when the COP counter expires? How does software prevent this from happening?

5. What is the purpose of the CPU interrupt vector table?

6. Why are microcontroller pins multiplexed between several I/O interfaces?

7. Why does a memory mapped I/O register typically behave differently from a RAM or ROM memory location?

8. Why is I/O synchronization required?

9. List the three I/O synchronization methods and briefly describe each. Which would be best used to integrate each of the following into a microcontroller system

    (a) Keypad

    (b) A device that requires 1us to complete a I/O write operation.

10. What is the principle advantage of using a device driver as opposed to directly accessing I/O interface registers from within a program?

11. Describe the difference between a blocking device driver subroutine and a non-blocking subroutine.

12. Write a sequence of instructions to configure pin PTA1 as an input pin with internal pull-ups enabled. The configuration of the other pins of Port A should not be affected.

13. Of the electrical specifications having to do with voltage levels on input/output pins

    (a) List and briefly describe the electrical specifications that apply to voltages on input pins, including both absolute maximum ratings and functional operating specifications?

    (b) List and briefly describe the electrical specifications that apply to voltages on output pins, including both absolute maximum ratings and functional operating specifications?

14. What is slew rate? Why do the GPIO pins include slew rate control? When would you want to disable slew rate control?

15. PTB1 is configured for low drive strength and is sourcing a load with a load current of 15 mA. PTB1 is also connected back to input pin PTA1.

    (a) When PTB1 is outputting a logic low, does PTA1 have a valid logic input on it?

    (b) When PTB1 is outputting a logic high, does PTA1 have a valid logic input on it?

16. Suppose a GPIO pin is driving a simple resistive load. In the sourcing configuration, what are the minimum and maximum values of the load resistance, assuming $V_{DD} = 3.0V$ and V_{OH} and V_{OL} must be within .5V of the power supply rails.

17. What are the minimum and maximum values for the resistor used in a switch interface circuit using the pull-down circuit configuration. Assume the resistor is a 1/4 Watt resistor and that a safety factor of 50% is desired i.e., the power dissipated in the resistor should not exceed 2/3 the rated value.

18. A 20 kOhm external pull-down resistor is connected to input pin PTB1. The internal pull-up is also enabled. What is the voltage on the input PTB1 (assume $V_{DD} = 3.0V$)?

19. PTB1 is configured as input. The internal pull-up resistor is not enabled and no external circuit is connected to pin PTB1. What is the input logic level on PTB1?

20. Suppose the normally-open switch assumed in Code Listing 1.2 is changed to a normally-closed switch. Nothing else is changed. Modify the switch driver such that application programs using the switch driver would still function as expected.

21. For the switch interface in Code Listing 1.2, it is desired to modify the hardware to use an external pull-down resistor with the normally-open pushbutton switch. Modify the driver so the hardware change does not affect the main program.

22. Compute a value for the current limiting resistor for the active-high LED circuit configuration connected to PTB0. Assume $V_f = 1.0V$, a diode current of 5mA is desired to get adequate brightness and $V_{dd} = 3.0V$.

23. Modify the driver in Code Listing 1.4 to accommodate a LED in the active-high configuration.

24. What would happen if an interrupt is not acknowledged within the ISR that services it?

25. Suppose a global variable is shared between the main program and an interrupt service routine. Why must interrupts be disabled before the main program modifies the variable?

26. What is the IRQ pin used for?

27. How is using a non-interrupt based driver for the IRQ pin different than polling pin PTA0? What is the advantage of each?

28. Why is the global I mask bit in the CCR set by the CPU upon entering an ISR?

29. Why can't CPU registers be used to exchange data between the main program and an ISR?

30. Modify the circuit in Figure 1.12, assuming a pull-down resistor configuration is used.

CHAPTER 2

Analog Input

Analog inputs must be encoded into digital form before they can be processed by a microcomputer system. This encoding process is called analog-to-digital conversion. The MC9S08QG4/8 microcontroller has both an analog-to-digital converter module, which coverts analog inputs to an 8 or 10 bit digital representation, as well as an analog comparator module that has a Boolean output that is triggered when the voltage on one input exceeds the other.

2.1 ANALOG TO DIGITAL CONVERSION

An analog-to-digital converter (ADC) takes an analog input voltage and produces a binary output representing its value. In a linear ADC, the range of analog input values is divided into equally space intervals, each of which is assigned a binary codeword, usually represented in an unsigned or two's complement code. The codeword that is output depends on the range within which the analog input signal falls.

2.1.1 ADC BASICS

The operation of a typical ADC is illustrated in Figure 2.1. Generally, the range of input voltages converted by the ADC is provided as two reference voltage inputs: the high reference voltage V_{RH} and the low reference voltage V_{RL}. The resolution of the ADC is the number of bits in the output codeword; an ADC with resolution n is called an n-bit ADC; thus, in the Figure 2.1, the ADC produces a 3b output and is therefore a 3-bit ADC. As shown in the figure, the analog input range is divided into equally spaced intervals each having size $\Delta = |V_{RH} - V_{RL}| / 2^n$. The parameter Δ represents the analog (or voltage) resolution; that is, the largest change that must be made in the analog input to change the binary output by ± 1.

During its operation, the output of the ADC encodes the number of Δ increments that the input voltage V_A is from V_{RL}. The integer value of the output is given by the formula $\text{output} = \text{ROUND}\left((V_A - V_{RL})/\Delta\right)$; $V_{RL} \leq V_A < V_{RH} - \Delta$. The use of the ROUND approximation in the ADC is the reason that the first ADC interval has width $.5\Delta$ and the last interval has width 1.5Δ; all other intervals have equal width Δ. Note that the input voltage can only be determined from the ADC output to within $.5\Delta$ of its actual value. Given the ADC output M, the input can be approximated by $V_A \approx V_{RL} + M\Delta = V_{RL} + \frac{M}{2^n}|V_{RH} - V_{RL}|$, which interprets the output as being in the middle of each of the equally space intervals, at the start of the first interval and $.5\Delta$ from the start of the last interval, as indicated by the blue points in Figure 2.1.

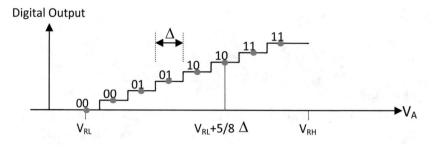

Figure 2.1: Input-output relationship of an analog-to-digital converter.

Example 2.1. A 5bit ADC has V_{RL} tied to 1V and V_{RH} tied to 6V. If the analog input is $V_A = 2.32$V, what is the binary output of the ADC?
Solution:

$$\text{output} = \text{ROUND}\left(\frac{V_A - V_{RL}}{\Delta}\right) = \text{ROUND}\left(\frac{(V_A - V_{RL})\,2^n}{V_{RH} - V_{RL}}\right) = \text{ROUND}\left(\frac{(2.32 - 1)\,2^5}{6 - 1}\right) = 8.$$

Answer: the binary output is 01000.

Example 2.2. An 8bit ADC with V_{RL} tied to ground and V_{RH} tied to 3V shows a binary output of 01001011. What is an approximate value for the input voltage V_A? Within what range of voltages does the input value fall?
Solution: $\Delta = |V_{RH} - V_{RL}| / 2^n = 3/256$. An approximate value for the input voltage is $V_A \approx V_{RL} + M\Delta = 0 + 75\,(3/256) = 0.88$V. The range of values is approximately $74.5\Delta \le V_A < 75.5\Delta$, where $\Delta = 3/256$.
Answer: $V_A \cong 0.88$V and 0.873V $\le V_A < 0.885$V.

2.1.2 CONVERTING ADC OUTPUT TO FIXED-POINT

Given the binary output M of the ADC, the input voltage V_A can be approximated as the number of Δincrements that it is from the low reference voltage, V_{RL}. For the MC9S08QG4/8 ADC, $V_{RL} = V_{SS} = 0$V, $V_{RH} = V_{DD}$ and the ADC precision is either $n = 8$ or $n = 10$. Thus, we have $V_A \approx V_{RL} + M\Delta = M \times \frac{V_{DD}}{2^n} = \frac{M}{2^n} \times V_{DD}$. To obtain a fixed-point approximation for V_A, the fixed-point constant representing V_{DD} needs to be multiplied by the 8b or 10b integer ADC output M divided by 2^n.

However, the division by 2^n is not explicitly required since it represents a shift in the binary point of V_{DD}. In addition, the fixed-point value $\frac{M}{2^n}$ has no integer part (since M is an n bit value) and the ADC output M can simply be interpreted as a fixed point value with the binary point assumed to be to the left of the most significant bit. A fixed point representation of V_A can thus be found by multiplying the integer M by V_{DD} and interpreting the result as having $n + v$ binary places, where v is the number of binary places in the fixed point representation of V_{DD}. This operation is illustrated in Figure 2.2 for an 8 bit ADC output. If the number of binary places in the constant V_{DD} is made equal to the number of binary places desired in the result, then the MSB of the product can simply be takes as the result, as shown in the figure and the following example.

Example 2.3.   Write a sequence of instructions that calls GET_ADC, converts the output to an 8b fixed point approximation for the ADC input voltage with 5 binary places, and stores the result to variable V_A. Assume $V_{DD} = 3.3V$?

Solution: The output M of GET_ADC is assumed to be a fixed point value with 8 binary places (i.e., a fraction of 256^{ths}). Since an 8b product is required to have 5 binary places, V_{DD} should be represented with a fixed point constant with 5 binary places; the MSB of the product will therefore contain the answer.

Answer:
```
1   VDD   equ   %01101010   ;3.3, as fixed pt. rounded to 5 places
2         bsr   GET_ADC     ;call to get ADC output in accumulator A
3         ldx   #VDD        ;load VDD for multiply
4         mul               ;product in X:A
5         stx   VA          ;store result (MSB of product, in X)
```

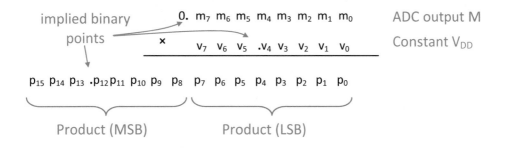

Figure 2.2: Conversion of ADC output to a fixed-point voltage.

The approximation $V_A \approx V_{RL} + M\Delta$ produces output voltage approximations of $V_{RL} \leq V_A \leq V_{RH} - \Delta$, which is $0 \leq V_A \leq V_{DD} - \Delta$; in other words, the lowest ADC output is interpreted as 0V while the highest is interpreted as $\frac{2^n-1}{2^n}V_{DD}$. In some applications, it may be more desirable or convenient to approximate the output voltage in the range $0 \leq V_A \leq V_{DD}$. This can be achieved with the approximation $V_A \approx \frac{M}{2^n-1} \times V_{DD}$, which results in the lowest ADC output being interpreted as 0V and the highest as V_{DD}. For example, using an 8 bit ADC the input voltage approximations would be $0, \frac{V_{DD}}{255}, \frac{2 \cdot V_{DD}}{255}, \frac{3 \cdot V_{DD}}{255}, \cdots, V_{DD}$. However, since the denominator is no longer a power of 2, a division must be explicitly performed.

Example 2.4. Repeat Example 2.3 such that the approximation for V_A is in the range $0 \leq V_A \leq V_{DD}$.

Solution: The output M of GET_ADC is multiplied by V_{DD}. The 16bit product is moved into H:A to be divided by 255.

Answer:

```
1   VDD    equ    %01101010    ;3.3, as fixed pt. rounded to 5 places
2          bsr    GET_ADC      ;get ADC output in accumulator A
3          ldx    #VDD         ;load VDD for multiply
4          mul                 ;product in X:A (from multiply)
5          pshx
6          pulh                ;product in H:A (for divide)
7          ldx    #255         ;load divisor
8          div                 ;divide
9          cphx   #$80FF       ;compare remainder with .5
                                  (X is $FF from above)
10         blo    DONE         ;if less, there is no rounding
11         inca                ;round up
12  DONE   sta    VA           ;store result (msB of product, in X)
```

2.2   ADC ON THE MC9S08QG4/8

The MC9S08QG4/8 contains a 10 bit ADC that has 32 multiplexed input channels, 12 of which are connected to internal or external sources; however the ADC can convert the voltage on only one of these sources at a given time. Up to eight external analog inputs can be sampled from pins ADP0 through ADP7. Four internal sources allow sampling V_{DD} or V_{SS}, a constant 1.2V bandgap reference voltage and an internal temperature sensor. The ADC reference voltages are internally fixed to $V_{RL} = V_{SS}$ and $V_{RH} = V_{DD}$. The input voltage must be kept to within the functional limits defined for I/O pins on the data sheet; if the voltage can exceed V_{DD} or V_{SS}, it should be conditioned. The ADC can also be configured to produce 8 bit outputs when 10 bit precision is not required for the application.

The ADC on the MC9S08QG4/8 is a linear, successive approximation ADC. A successive approximation ADC requires multiple clock cycles to convert the analog input to a digital output. Thus, when a conversion is performed, software must wait until the conversion is complete before taking the output. Interrupt synchronization, delay synchronization or polling can be used to synchronize software with the ADC. Because $V_{RH} = V_{DD}$ and $V_{RL} = V_{SS}$, the ADC inputs connected to V_{DD} and V_{SS} are for ADC testing purposes, since the ADC output should always be \$FF and \$00, respectively, for these inputs.

2.2.1   MC9S08QG4/8 ADC I/O INTERFACE REGISTERS

There are 8 memory-mapped registers associated with the ADC on the MC9S08QG4/8: two status and control registers, two data registers and four control registers. The control registers control the primary operating mode of the ADC, such as selecting the channel to convert, configuring the operating mode, and enabling microcontroller pins as inputs. The status registers report back status about the progress of the conversion. The data registers contain the ADC output when a conversion is complete.

The first ADC Status and Control register (ADCSC1) is used to configure the ADC and contains a single status flag. The format of this register is shown Figure 2.3. The 5 bit ADCH field selects the ADC channel to be sampled. Writing a value to the ADC status and control register initiates a new conversion. Table 2.1 shows the channel assignments for the MC9S08QG4/8 ADC. When the ADCH is set to 11111 (channel 31) the ADC is disabled. Channels 0 through 7 correspond to the eight pin inputs ADP0 through ADP7, respectively. To use the microcontroller pins, the individual pin enable must be set in the APCTL1 register. The format of this register is 8 bits, one per pin, that when set enable the ADC function on the given pin. An internal temperature sensor is connected to ADC channel 26; this has a coarse resolution and is intended for estimating chip temperature for parameter compensation rather than measuring ambient temperature. The internal bandgap reference is a fixed internal voltage reference that remains nearly constant, independent of the temperature of the device or V_{DD}. The nominal value of this reference is 1.2V. To use this as an ADC input, the bandgap reference voltage must be enabled in the System Power Management Status and Control register.

ADCO configures the conversion mode of operation to be either single conversion (ADCO=0) or continuous conversion (ADCO=1). With single conversion mode, the input voltage on the chosen pin is converted by the ADC, after which the ADC stops until another conversion is initiated by software. In continuous conversion mode, the ADC performs the conversion continually on a single ADC pin, initiating a new conversion on that same pin as soon as the previous one completes. Thus, continuous conversion mode is useful if the software requires repeated conversions on only one of the ADC pins. The AIEN bit is the ADC interrupt enable bit; when AIEN=1, an interrupt request is raised at the completion of each conversion cycle. The conversion complete (COCO) bit is a status flag that is set by the ADC upon completion of a conversion operation (indicating that the data in the ADC data register is valid). This flag is cleared automatically when software reads from the ADC data register or writes to ADSC1.

ADCSC1: ADC Status & Control Register 1 (memory mapped at address $0010)

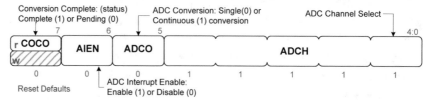

ADCSC2 ADC Status & Control Register 2 (memory mapped at address $0011)

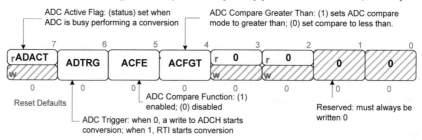

Figure 2.3: MC9S08QG4/8 ADC status and control registers.

Table 2.1: ADC Input Channel Assignments.

ADC Channel	ADC Input
00000-00111	Pins ADP0 through ADP7
01000-10101	Tied to V_{SS}
10110-11001	Reserved
11010	Temperature Sensor
11011	Internal Bandgap
11100	Reserved
11101	V_{DD}
11110	V_{SS}
11111	None (disables ADC)

The second ADC status and control register ADSC2 contains one status bit and three additional configuration options. Status flag ADACT becomes set while a conversion is in progress; this flag can be tested by software to determine if the ADC is currently in use. ADTRG is the conversion trigger select bit. When clear, an ADC conversion is initiated upon writing a valid channel value to ADSCR1. When set, a hardware trigger from the real time interrupt module initiates a conversion; this can be used to achieve a coarse periodic sampling.

The remaining two configuration bits are used to enable and configure the ADC compare function. When the compare function is enabled, COCO is not set and no conversion data is written to the ADC data registers unless the converted value is exceeds a threshold, either greater than or less than depending on the operating mode. Compare mode could be used, for example, to trigger an interrupt when the analog voltage on a pin exceeds a certain threshold. When set, AFCE enables the compare function. When compare is enabled, ACFGT selects whether the comparison is true when the voltage is less than the threshold (ACFGT=0) or greater than or equal to the threshold (ACFGT=1). The threshold value is programmed into the compare value high (ADCCVH) and low (ADCCVL) registers.

ADCFG: ADC Configuration Register (memory-mapped at address: $0016)

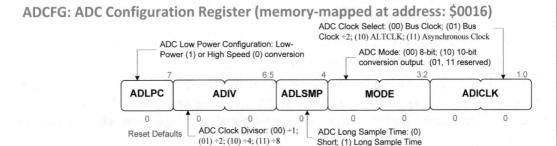

Figure 2.4: ADC configuration register format.

The ADC Configuration Register (ADCCFG) allows additional control over the configuration of the ADC. When set, ADLPC enables low power conversion at the expense of conversion time. When fastest conversion is desired, the bit should be cleared. The two bit mode field configures 8 or 10 bit conversion; the only valid values for this field are 00 (8 bit conversion) or 10 (10 bit conversion). ADICLK selects which clock source clocks the ADC. Valid choices are: 00, the bus clock; 01, the bus clock divided by 2; or 10 asynchronous clock within the ADC that can vary from approximately .4 MHz to 4 MHz. ADIV is the ADC clock divider select. The ADC conversion clock frequency is the ADC input clock frequency divided by 2^{ADIV}, where ADIV is treated as a 2 bit unsigned integer. In low power mode, the ADC clock must be divided to fall within .4 and 4 MHz; when not in low power mode, this range increases to 8 MHz.

The ADC result registers are read-only registers that hold the last converted output of the ADC. The format of these registers is shown in Figure 2.5. For 8 bit ADC mode, only ADCRL is used; in 10 bit mode the most significant 2 bits are stored in the least significant bit positions of ADCRH and the least significant 8 bits are in ADCRL. The data registers are interlocked in 10 bit mode, meaning that after a conversion a read of ADCRH prevents the ADC from overwriting the value in the registers until ADCRL is subsequently read. This prevents the situation where a new conversion completes between loads from ADCRH and ADCRL, which could lead to invalid results. Thus, in 10b mode ADCRH should always be read before ADCRL. In both modes, a read from ADCRL automatically clears the conversion complete flag (COCO) in ADCSC1.

ADCRH: ADC Result High (Address: $0012)

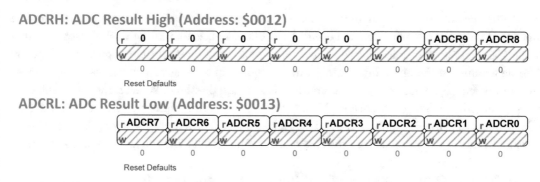

Figure 2.5: Format of the ADC result registers.

Figure 2.6 summarizes the configuration options for the MC9S08QG4/8 microcontrollers and the order in which these steps should be performed. It is required that the ADCSC1 register be configured last, since writing to it initiates the conversion and all configuration options must be set prior to this.

2.3   DRIVER EXAMPLES FOR THE MC9S08QG4/8 ADC

In this section, six drivers for the ADC are presented to illustrate the wide variety of ways the MC9S08QG4/8 ADC can be used. The ADC can be used in 8 or 10 bit mode, can perform continuous or single conversion, can operate with interrupt synchronization or polling and can be used with the compare function enabled or disabled. Thus, the drivers in this section comprise only a small fraction of the ways in which the ADC can be applied.

Code Listing 2.1 lists common equate constants used in the code examples in this section, as well as two shared subroutines. Including this code once here allows shortening of the subsequent driver examples; however, it must be included with each of the drivers for them to work correctly.

Lines 1 through 8 define the memory-map locations of the ADC status, control and data registers. The constant STOPADC on line 9 defines the value that, when written to the ADCSC1 register, powers off the ADC. COCO, ADACT, AIEN and ADCO define the bit positions of useful flags and configuration bits in their respective registers.

PINMASK and ADCPIN are used by the drivers to determine which microcontroller pin or pins will be used by the driver. When a driver uses a single pin, ADCPIN should be set to the number (0-7) corresponding to the pin, ADP0 through ADP7, to be used. When multiple pins are used by a driver, PINMASK defines a bit-mask in which a 1 in bit position i corresponds to enabling pin PADi. For example, setting PINMASK to $13 corresponds to the driver using pins PAD0, PAD1 and PAD4.

For drivers that convert the ADC output to a fixed-point voltage, V_{DD} is a fixed point constant representing the power supply voltage. It is important that the value of V_{DD} accurately reflect the power supply voltage to achieve accurate computation. The number of binary places in V_{DD} can be

Step	Summary	Details		
Step 1	Configure 8 bit or 10 bit mode and clock source in ADCCFG	Bit 7	0 (high speed)	1 (low power)
		Bits 6:5	00 (÷1 clock); 01 (÷2 clock); 10 (÷4 clock); 11 (÷8 clock)	
		Bit 4	0 (short sample time)	1 (long sample time)
		Bits 3:2	00 (8 bit)	10 (10 bit conversions)
		Bits 1:0	00 (BUS clock)	01 (BUS clock ÷2)
			11 (Local asynchronous ADC clock)	
Step 2	Configure conversion trigger and compare function in ADCSC2	Bit 7	n/a (status flag, always write 0)	
		Bit 6	0 (software trigger); 1 (hardware trigger from RTI)	
		Bit 5	0 (disable compare)	1 (enable compare)
		Bit 4	0 (compare less than)	1 (compare greater than)
		Bits 3:0	0000 (always 0)	
Step 3	Enable ADC pin(s) in APCTL1	Bit i	0 (ADC function on pin ADi disabled)	
			1 (ADC function on pin ADi enabled)	
Step 4	Configure compare trigger value in ADCCVH:ADCCVL if using compare	Bits 15:8	upper byte of compare value	
		Bits 7:0	lower byte of compare value	
Step 5	Configure operating mode and ADC source in ADCSC1	Bit 7	n/a (status flag)	
		Bit 6	0 (interrupts disabled)	1 (interrupts enabled)
		Bit 5	0 (single conversion)	1 (continuous conversion)
		Bit 4:0	channel number to convert	

Figure 2.6: Steps required for configuring the MC9S08QG4/8 ADC.

altered as needed and the number of binary places used in the converted output voltage will always be the same as the number used in V_{DD}. TRUE and FALSE are convenient constants used for testing and for assigning to Boolean variables.

Driver subroutine ADCSTOP, which begins on line 22, simply writes the value STOP_ADC ($1F) to ADCSC1, which sets the CH field to %11111, powering down the ADC. Although this subroutine is only a single instruction, it is easier for the embedded systems programmer to remember to call ADCSTOP than to remember the instruction needed to stop the ADC.

Driver subroutine CHKPINMASK takes a pin number, 0-7, as an argument in accumulator A and returns with the Z flag indicating the state of the bit in the PINMASK. This is used by two of the ADC drivers to verify that a pin has been configured for use. The subroutine computes X=1<<A on lines 29-33 with a simple counter loop. The value (1<<A) is then logically ANDed with the constant PINMASK, which results in Z=1 if the bit is zero and Z=0 otherwise.

```
1    ADCSC1      equ   $10          ;ADC status and control register 1
2    ADCSC2      equ   $11          ;ADC status and control register 2
3    ADCRH       equ   $12          ;ADC result register high byte
4    ADCRL       equ   $13          ;ADC result register low byte
5    ADCCVH      equ   $14
6    ADCCVL      equ   $15
7    ADCCFG      equ   $16          ;ADC configuration register
8    APCTL1      equ   $17          ;ADC pin control register
9    STOPADC     equ   %00011111    ;Mask value for ADCSC1 to power down ADC
10   COCO        equ   7            ;Bit position of COCO bit in ADCSC1
11   ADACT       equ   7            ;Bit position of ADACT bit in ADCSC2
12   AIEN        equ   6            ;Bit position of AIEN mask in ASCSC1
13   ADCO        equ   5            ;Bit position of ADCO bit in ADCSC1
14   PINMASK     equ   %11111111    ;Mask defining which ADC pins are used
15   ADCPIN      equ   0            ;Pin number for single pin ADC drivers
16   VDD         equ   %11000000    ;Constant VDD voltage, fixed point
17   TRUE        equ   1            ;Constant true
18   FALSE       equ   0            ;Constant false
19
20   ;--------------------ADC_STOP Subroutine--------------------------
21   ; power off the ADC by writing CH=11111 to ADCSC1
22   ADC_STOP    mov   #STOPADC,ADCSC1
23               rts
24   ;--------------------CHECKPIN Subroutine-------------------------
25   ;return Z=1 if bit i is clear (zero) in PINMASK, Z=1 otherwise
26   ;I passed in accumulator A
27   CHKPINMASK  pshx               ;callee save
28               psha
29               ldx   #$01         ;onemask=%00000001
30   LOOPMASK    cmp   #0           ;while I is not zero
31               beq   ENDLPMASK
32               lslx               ;onemask = onemask << 1
33               dbnza LOOPMASK     ;A=A-1; repeat loop
34   ENDLPMASK   txa                ;copy one-mask into A
35               bit   #PINMASK     ;test bit I in PINMASK, Z=1 if clear
36               pula               ;callee restore
37               pulx
38               rts
```

Code Listing 2.1: Equate Pseudo-Ops for ADC Driver Examples.

2.3.1   BASIC 8-BIT SINGLE-PIN POLLED I/O DRIVER

Code Listing 2.2 is an example of an 8 bit polled I/O single pin driver for the MC9S08QG4/8 ADC. This driver is useful when a single ADC input is used in an embedded system. The necessary equates from Code Listing 2.1 must be placed in the appropriate place in the final program. There are no global variables used by this ADC driver.

The driver initialization routine must be called from within the main program section where the microcontroller is configured. This routine, INIT_ADC, sets up the necessary ADC control registers.

ADCSC1 is not configured in the initialization subroutine for this driver because a write to it initiates a conversion; instead, it is configured in driver subroutine GET_ADC to start a new conversion. The initialization routine first calls driver subroutine ADC_STOP to ensure that the ADC is in a stopped state; while turning off the ADC may not be required, having it in this state when not in use is a good idea for reduced power consumption and to ensure that if the ADC is already active, it will be deactivated before configuring. Initialization then proceeds with the first three steps specified in Figure 2.6. Specifically, ADCCFG is configured for high speed, no clock division, short sample time, 8 bit conversions, and the bus clock as the ADC clock input. Subsequently, $00 is written to ADCSC2 to disable the compare function and to configure the software trigger so that a write to the CH field in ADCSC1 initiates a new conversion. Finally, the selected ADC pin is enabled by setting the corresponding enable bit in APCTL1. The initialization routine then returns.

The purpose of driver subroutine GET_ADC is to initiate a conversion on the selected pin and return the ADC output in accumulator A. This subroutine is blocking; it does not return until the conversion is complete. On line 10, the ADC pin number is written to ADCSC1, which initiates a conversion on that pin; note that because the channel parameter is in the range 0-7, bits 5 and 6, which are written to ADCO and AIEN, remain 0. Thus, writing the channel number starts a single conversion operation on the selected channel and disables interrupts, which is the desired behavior for this driver. Line 11 is the ADC polling loop. The BRCLR instruction branches to itself as long as the COCO bit (bit 7) of the ADCSC1 is clear. When COCO is raised, indicating that the conversion is complete, the CPU will exit the polling loop and resume execution on line 12, at which point the ADC result register is read into A. The read also clears COCO as a side effect, ensuring that it will be clear the next time GET_ADC is called. ADC_STOP is then called before returning to the caller to ensure that the ADC is shut down until the next time GET_ADC is called.

```
 1   ;--------------------INIT_ADC Subroutine--------------------------
 2   INIT_ADC    bsr   ADC_STOP          ;make sure ADC is in stopped state
 3               mov   #$00,ADCCFG       ;fast clock settings,8b mode
 4               mov   #$00,ADCSC2       ;software trigger
 5               bset  ADCPIN,APCTL1     ;enable ADC pin
 6               rts
 7   ;--------------------GET_ADC Subroutine--------------------------
 8   ;perform a single conversion on ADC;
 9   ;  ADC data returned in index register A
10   GET_ADC     mov   #ADCPIN,ADCSC1    ;start conversion: write CH to ADCSCR
11   WAIT_ADC    brclr COCO,ADCSC1,WAIT_ADC ;poll until conversion complete
12               lda   ADCRL             ;read data, clearing COCO
13               bsr   ADC_STOP          ;stop (power down) the ADC
14   DONE_ADC    rts
```

Code Listing 2.2: 8-Bit Single Conversion ADC Driver Using Pin PAD0.

2.3.2 BASIC 10-BIT SINGLE CONVERSION WITH SOFTWARE SELECTABLE PIN

Code Listing 2.3 illustrates a polled 10b single conversion driver for the ADC. This driver performs a 10b conversion on any of the eight ADC pins; the specified pin is passed as a parameter to driver subroutine GET_ADC. This driver could be used to occasionally sense the analog voltages on several microcontroller pins, for example, such as when multiple analog sensors or switches are interfaced.

The initialization routine is similar to that of Code Listing 2.2. The main differences include that ADCCFG is configured for 10 bit conversions and multiple ADC pins are enabled on the microcontroller by writing PINMASK to APCTL1. Recall that PINMASK is an 8 bit one-mask that defines which microcontroller pins will be used as analog inputs in the application. If all ADC pins were simply enabled, the lower priority functions on these pins would cease to function and could not be used for other purposes.

The purpose of driver subroutine GET_ADC is to initiate a conversion on the channel number (0-7) passed in accumulator A and return the ADC output in HX. In addition, the carry flag is set if an error is encountered and cleared otherwise. There are two cases that could cause an error: the channel number is not valid (not in the range 0-7) or the ADC pin corresponding to the selected channel has not been enabled. This subroutine is blocking; it does not return until the conversion is complete. On lines 12 and 13, the channel parameter in accumulator A is checked to ensure that it is valid (not higher than 7); if this is not true, the carry flag is set (line 34) to indicate error and the subroutine simply returns. On lines 14 and 15, the channel is checked to determine if its corresponding pin has been enabled for use with the ADC. Recall that the subroutine CHKPINMASK returns Z=0 if PINMASK&(1<<A) is true, indicating that the bit corresponding to the channel in A is set in PINMASK; if not, the pin has not been enabled for use with the ADC in INIT_ADC. If the channel number is valid, the channel number is written to ADCSC1 on line 16, which initiates a conversion. Note that because the channel parameter is in the range 0-7, bits 5 and 6, which are written to ADCO and AIEN, remain 0. Thus, writing the channel number starts a single conversion operation on the specified channel with interrupts disabled, which is the desired behavior for this driver.

Line 17 is the ADC polling loop, which is similar to the polling loop in Code Listing 2.2. When the ADC completes the conversion, the ADC result registers are subsequently read into HX (Line 44). Note that ADCRH and ADCRL are arranged in memory in big-endian order, allowing the entire 16 bit value (zero-extended 10 bit ADC result) to be loaded into HX with a single load operation. This read also clears COCO as a side-effect. After calling ADC_STOP to shut down the ADC, the carry flag is cleared before exit to indicate there were no errors.

2.3.3 INTERRUPT-BASED 8-BIT DRIVER

Code Listing 2.4 illustrates the use of interrupt synchronization and continuous conversion mode to provide repeated conversion of a single ADC pin with automatic conversion of the ADC output to a fixed-point voltage value. This allows the software to always have access to the latest value of the voltage on the selected pin, without having to initiate a conversion, poll and convert the ADC output

```
 1    ;--------------------INIT_ADC Subroutine----------------------
 2    INIT_ADC  bsr    ADC_STOP   ;
 3              mov    #$08,ADCCFG      ;fast clock settings,10b mode
 4              mov    #$00,ADCSC2      ;software trigger
 5              mov    #PINMASK,APCTL1 ;enable all ADC pins
 6              rts
 7    ;--------------------GET_ADC Subroutine-----------------------
 8    ;perform a single conversion on ADC;
 9    ;   channel number (0-7) passed in accumulator A
10    ;   ADC data returned in index register HX
11    ;   carry flag indicates error (i.e. channel number not 0-3)
12    GET_ADC   cmp    #7        ;ensure channel between 0 and 7
13              bhi    ADC_ERR   ;if not, goto exit
14              bsr    CHKPINMASK
15              beq    ADC_ERR   ;Z=1 indicates pin is not in PINMASK
16              sta    ADCSC1    ;start conversion: write CH to ADCSCR
17    WAIT_ADC  brclr  COCO,ADCSC1,WAIT_ADC  ;wait for conversion
18              ldhx   ADCRH     ;read data, clearing COCO
19              bsr    ADC_STOP  ;stop (power down) the ADC
20              clc              ;return value C=0 (no error)
21              bra    DONE_ADC
22    ADC_ERR   sec              ;return value C=1 (error)
23    DONE_ADC  rts
```

Code Listing 2.3: 10 Bit Single Conversion ADC Driver with Selectable Channel.

to a voltage. As in the previous examples, the required equates are listed in Code Listing 2.1. Recall from Section 2.1.1 that converting the ADC output to a fixed-point voltage requires multiplying by V_{DD} and discarding the least significant byte of the product (or using it to round). Thus, V_{DD} should be defined to have the same number of binary places desired in the converted voltage output. In Code Listing 2.1, V_{DD} is shown to be 3.0 volts, represented with 6 binary places (%11000000).

Since this driver is interrupt-based, it must exchange data with the main program via global variables. Two global variables are used: ADCVOLTAGE, which holds the value of the ADC output converted to a fixed-point voltage, and ADCCOCO, which is a Boolean variable that is set by the driver each time a new value is written to ADCVOLTAGE. Software can poll ADCCOCO, just like the COCO bit in ADCSC1, to determine if a new voltage value is in ADCVOLTAGE. As written, the driver requires that these global variables will be defined in the zero page; if not, the driver code should be modified to not use MOV instructions to access them. Note that since continuous conversion and interrupts are enabled, the driver can write to these variables at any time. As described in Section 1.4, this implies that access to these variables should be atomic. Example 2.5 illustrates how this is done.

The initialization subroutine INIT_ADC configures the operating mode of the ADC and initiates the continuous conversion of the selected pin. Line 5 stops the ADC to ensure it is not active while configuration registers are being modified. Lines 6 and 7 configure the ADC to operate in 8 bit conversion mode, slow clock settings, software trigger and compare function disabled. Slow clock

```
1    ; ---------------Required Global Variables--------------------
2    ;ADCVOLTAGE  ds.b 1              ;ADCOUTPUT converted to (fixed pt.) volts
3    ;ADCCOCO     ds.b 1              ;set TRUE after each write to ADCVOLTAGE
4    ;-------------------INIT_ADC Subroutine----------------------
5    INIT_ADC     bsr  ADC_STOP       ;disable ADC while configuring
6                 mov  #$F1,ADCCFG     ;slow clock settings,8b mode
7                 mov  #$00,ADCSC2     ;software trigger, no compare
8                 bset ADCPIN,APCTL1   ;enable selected ADC pin
9                 mov  #FALSE,ADCCOCO  ;initialize driver global variable
10               ;start continuous conversion on ADCPIN with interrupts
11                mov  #(ADCPIN|(1<<AIEN)|(1<<ADCO)),ADCSC1
12                rts
13   ;-------------------ADC ISR---------------------------------
14   ADCISR       lda  ADCRL          ;clear COCO, ACK interrupt
15                ldx  VDD            ;multiply by fixed point rep. of VDD
16                mul
17                stx  ADCVOLTAGE     ;MSB of product is fixed point voltage
18                mov  #TRUE,ADCCOCO  ;set ADCCOCO true
19                rti
```

Code Listing 2.4: Interrupt Driver for 8 Bit ADC with Automatic Voltage Conversion.

settings were selected to minimize the frequency of interrupts; however, fast clock settings may be unavoidable depending on the characteristics of the analog signal being sampled. The selected ADC pin is enabled with a BSET instruction on line 8. Global variable ADCCOCO is initialized to FALSE on line 9, ensuring that the main program will not read an uninitialized value in ADCVOLTAGE. Line 11 begins the continuous conversion. The expression (ADCPIN|(1<<AIEN)|(1<<ADCO)) is an assembler constant expression that creates the necessary configuration value for ADCSC1 that will set the CH field to ADCPIN and set the ADCO and AIEN bits. (1<<AIEN) results in the binary value %01000000, which is the mask for the AIEN bit in ADCSC1. Likewise, (1<<ADCO) results in a one-mask for the ADCO bit: %00100000. The value of ADCPIN is logically OR'ed with these masks by the assembler, forming the correct constant value to write to ADCSC1. This write begins a continuous conversion on pin ADCPIN; each time a conversion completes, an ADC interrupt request will be made to the CPU and another conversion automatically starts.

The interrupt service routine, ADCISR, acknowledges the interrupt request by clearing COCO, which is done as a side-effect of loading the ADC output from ADCRL. To convert this to a voltage, the value is multiplied by V_{DD} and the MSB of the result is stored in global variable ADCVOLTAGE. Subsequently, ADCCOCO is set to true to convey to software that a new value is present in ADCVOLTAGE. If software did not read the last value in ADCVOLTAGE, it is overwritten; this is similar to the behavior of the ADC result registers when continuous conversion is enabled. If it is crucial that each ADC output be maintained, an array of samples would have to be buffered.

Example 2.5. Write the instructions necessary to load into accumulator A the ADC output produced by the ISR in Code Listing 2.4.

Solution: The following code illustrates a polling loop that, as long as ADCCOCO is 1, reads the value in ADCVOLTAGE and clears ADCCOCO so that the next access will retrieve a new voltage value.

Answer: Type the Solution.

```
1  WAITADCCOCO   tst  ADCCOCO          ;poll ADCCOCO
2                beq  WAITADCCOCO      ;until new value ready (not zero)
3                sei                   ;disable interrupts
4                lda  ADCVOLTAGE       ;get new value
5                mov  #FALSE,ADCCOCO   ;ADCCOCO=FALSE for next access
6                cli                   ;re-enable interrupts
```

2.3.4 MULTIPLE ADC PIN SCANNING USING INTERRUPT-BASED DRIVER

In Code Listing 2.4, interrupt synchronization and continuous conversion are used to achieve repeated conversion of a single ADC input pin. In some applications, multiple analog input devices are required to be integrated into the microcontroller system. The interrupt service routine can be made to process the ADC output that triggered the interrupt request, then begin a conversion on another ADC channel. This channel scanning approach is illustrated in Code Listing 2.5. The driver uses PINMASK to define which ADC pins will be scanned and ADCPIN to define the first pin from PINMASK scanned. The pins are scanned in round-robin fashion, from right to left in PINMASK; after each complete pass through all channels, a global variable SCANCOCO is set to TRUE to indicate that the values are ready. The ADC outputs are stored, in unmodified form, in a global array called ADCSCANARRAY, where ADCSCANARRAY[i] holds the ADC output forPIN PADi. This array is 8 bytes long and only contains valid values for those ADC pins listed in PINMASK. One additional 8 bit global variable, ADCSCANCHAN, holds the number of the ADC pin that is currently being converted. Continuous conversion mode, though not strictly required in this driver, is enabled to ensure that in case a conversion is somehow missed, the ADC will continue to operate and generate interrupt requests.

The driver initialization subroutine, INIT_ADC, stops the ADC and configures ADCCFG and ADCSC2. Since this is a multiple PIN driver, PINMASK defines all pins that will be used by the ADC; thus, PINMASK is written directly to APCTL1 on line 10 to enable the pins for use with the ADC. ADCPIN defines the first pin that is converted; this is the initial value written to global variable ADCSCANCHAN on line 11. Line 12 initiates the first conversion. The syntax used to generate the ADCSC1 configuration data is the same as that used in Code Listing 2.4. Before returning, global variable SCANCOCO is initialized to FALSE (line 13).

The interrupt service routine for this driver begins on line 16. H is first stacked since the ISR uses index register HX and the CPU does not automatically stack H. Following this, the ADC output is

```
1    ; ----------------Required Global Variables-------------------
2    ;all are to be defined in direct page
3    ;ADCSCANCHAN   ds.b   1           ;pin number of ADC pin being converted
4    ;ADCSCANARRAY  ds.b   8           ;array of ADC outputs, one for each pin
5    ;SCANCOCO      ds.b   1           ;set TRUE at end of each full scan
6    ;--------------------INIT_ADC Subroutine---------------------
7    INIT_ADC  bsr   ADC_STOP      ;stop ADC before configuring
8              mov   #$F1,ADCCFG   ;slowest clock settings,8b mode
9              mov   #$00,ADCSC2   ;software trigger, no compare
10             mov   #PINMASK,APCTL1 ;enable scanned pins ADPi
11             mov   #ADCPIN,ADCSCANCHAN ;set initial channel number
12             mov   #(ADCPIN|(1<<AIEN)|(1<<ADCO)),ADCSC1 ;start conversion
13             mov   #FALSE,SCANCOCO ;clear the SCANCOCO global flag
14             rts
15   ;--------------------ADC ISR---------------------------------
16   ADCISR    pshh                ;callee save h
17             lda   ADCRL         ;get ADC output (clear COCO, ACK interrupt
18             ldx   ADCSCANCHAN   ;get current channel
19             clrh                ;zero extend into hx
20             sta   ADCSCANARRAY,X ;ADCSCANARRAY[channel]=ADC output
21             txa                 ;channel into A
22   ADCPINLP  inca                ;increment channel to next channel
23             and   #$07          ;mask for modulo 8 addition (8 becomes 0)
24             bsr   CHKPINMASK    ;check if channel is in pinmask
25             beq   ADCPINLP      ;repeat until we find a channel in mask
26             sta   ADCSCANCHAN   ;update current channel variable
27             cmp   #ADCPIN       ;are we back to first pin again
28             bne   ADCNXTSCN
29             mov   #TRUE,SCANCOCO ;if yes, then set SCANCOCO
30   ADCNXTSCN ora   #((1<<AIEN)|(1<<ADCO)) ;form mask with AIEN and ADCO
31             sta   ADCSC1        ;start next conversion
32             pulh                ;callee restore
33             rti                 ;return to interrupted code
```

Code Listing 2.5: Scanning Multiple ADC Pins with Interrupt Driver.

loaded on line 17, which also acknowledges the interrupt request by clearing the COCO bit in ADCSC1. This value needs to be stored to array location ADCSCANARRAY[ADCSCANCHAN]; lines 18-21 perform the store to the array.

Beginning on line 21, the remaining function of the ISR is to determine the next pin the ADC needs to convert. A do-until loop (lines 21 through 25), is used to find the next ADC pin for which there is a 1 in the PINMASK. Since valid pin numbers are 0-7, the current pin number, which is in accumulator A, must be incremented modulo-8; that is, A=(A+1) mod 8. The modulo is accomplished with the AND operation on line 23. After each increment, CHKPINMASK is called to determine if the corresponding bit in PINMASK is 1. Until a 1 is found, the loop repeats. Upon falling out of the loop, the new pin number is written to the global variable ADCSCANCHAN (line 26). Next, the pin is compared with the first pin scanned, ADCPIN, to determine if a complete pass has been made through all channels. If so, SCANCOCO is set to TRUE (line 29).

On line 30, the value to be written to ADCSC1 is composed. An OR-mask operation is used to set the bits corresponding to AIEN and ADCO in A (line 30) before the value is written to ADCSC1 to begin the next conversion (line 31). At this point, register H is restored and control is returned to the interrupted code.

2.3.5   8-BIT POLLED DRIVER WITH COMPARE FUNCTION

Code Listing 2.6 illustrates a driver for the ADC using the compare function. Recall that when the compare function is enabled, the ADC output is compared to a threshold value and the conversion complete (COCO) flag becomes set only if the compare results in true. The compare can be configured as less-than or greater-than-or-equal. This driver can be used, for example, with a sensor to detect when the sensor output exceeds a threshold value (such as a proximity sensor indicating closeness to an obstacle). The driver includes a non-blocking subroutine that takes as a parameter and 8 bit compare value, and returns true (through the carry flag, C=1) if the voltage on pin ADP0 results in an ADC output that is greater than the compare value; otherwise, the carry flag is cleared. It also includes a similar blocking subroutine that returns only after the compare results in true.

The initialization subroutine is similar to the previous ADC drivers. The ADC is first disabled and then configured for 8 bit operation. The main difference is that on line 4, the value written to ADCSC2 has the compare function enable bit set as well as the compare greater than or equal bit.

The driver subroutine COMP_ADC compares the ADC output, corresponding to the analog voltage on pin PTA0, to the compare value passed as a parameter in accumulator A. The first instruction on line 12 writes the compare value to the ADC compare value low register. Since this driver is using the ADC in 8 bit mode, the compare value high register is not used. On line 13, the pin number is written to ADCSC1, starting the ADC conversion on the selected pin. Since the ADC only sets COCO in compare mode if the compare results in true, and this is a non-blocking driver, COCO cannot be polled to determine when the ADC conversion is complete. Instead, the ADACT (ADC conversion active) flag in ADCSC2 is polled on line 14 until it becomes zero, indicating that the conversion has completed. Though the subroutine does wait until the conversion operation is complete, the non-blocking behavior described is with respect to the compare being true. The BRCLR instruction on line 15 then tests COCO to determine if the compare resulted in true. If not, the branch jumps to DONE_ADC, which powers down the ADC before returning from the subroutine. If COCO is set, the load on line 16 gets the ADC output and puts it in accumulator A, clearing COCO as a side effect. Recall that the BRCLR instruction sets or clears the carry flag based on the value of the bit being tested. Thus, upon returning from the subroutine, the carry flag contains the value of the COCO bit after the conversion, which is true if the compare resulted in true and false otherwise.

The driver subroutine WAIT_COMP is similar to COMP_ADC, except it is a blocking subroutine that does not return until the voltage on the selected pin causes the compare to be true. Thus, the polling loop includes a reset of the watchdog timer (since there is no way to predict how long it will

```
1    ;---------------------INIT_ADC Subroutine---------------------------
2    INIT_ADC     bsr  ADC_STOP      ;stop ADC while configuring
3                 mov  #$00,ADCCFG    ;fast clock settings,8b mode
4                 mov  #$30,ADCSC2    ;software trigger, compare >= enabled
5                 bset ADCPIN,APCTL1 ;enable ADC pin
6                 rts
7    ;-------------------COMP_ADC Subroutine---------------------------
8    ;Initiate a capture on PAD pin;
9    ;  Capture value parameter passed in A
10   ;  Returns converted value in A and C set if compare true
11   ;  Returns C clear with A unmodified if compare false
12   COMP_ADC     sta  ADCCVL
13                mov  #ADCPIN,ADCSC1;start compare on pin ADCPIN
14   WAIT_ADC     brset ADACT,ADCSC2,WAIT_ADC ;wait for conversion
15                brclr COCO,ADCSC1,DONE_ADC
16                lda  ADCRL          ;else load converted value
17   DONE_ADC     bsr  ADC_STOP
18                rts
19   ;--------------------WAIT_COMP Subroutine---------------------------
20   ;Initiate a capture on PAD pin;
21   ;  Poll until capture is true
22   WAIT_COMP    sta  ADCCVL
23                mov  #ADCPIN,ADCSC1;start compare on pin ADCPIN
24   WAIT_ADC     sta  WATCHDOG       ;feed watchdog to prevent reset
25                brclr COCO,ADCSC1,WAIT_ADC ;wait for conversion
26                lda  ADCRL
27                bsr  ADC_STOP
28                rts
```

Code Listing 2.6: Polled I/O Driver for 8 Bit ADC Compare Function on Pin PAD0.

be until the compare is true), and polls COCO until it indicates that the compare is true. When the compare becomes true, control falls out of the poling loop, the ADC result is read to clear COCO and the subroutine powers down the ADC before returning. Upon return accumulator A contains the ADC output that caused the compare to become true.

2.3.6   8-BIT INTERRUPT-BASED DRIVER WITH COMPARE FUNCTION

Code Listing 2.7 illustrates the use of the ADC compare function with interrupt I/O synchronization. The driver is designed to monitor the value of a single ADC pin against a threshold; when the analog voltage falls below the threshold an interrupt request is made. Although the ISR is programmed to simply set a global variable to true (indicating that the compare event happened), it could be modified to perform any function required when the analog voltage falls below the threshold.

Except for the configuration of less than compare on line 6, the driver initialization subroutine is identical to Code Listing 2.6. Subroutine START_ADC is called by software to start a compare operation on pin ADCPIN (defined by an equate pseudo-op in Code Listing 2.1). The subroutine receives the compare value in accumulator A and writes it to the compare low register on line 11. On line 12, the driver global variable COMPARETRUE is initialized to FALSE. Line 13 starts the continuous conversion on pin ADCPIN with interrupts and continuous conversion enabled. This same instruction was used in Code Listing 2.4 to initiate a conversion. Once the conversion is started, the subroutine returns to the calling program.

```
1    ; ---------------Required Global Variables----------------------
2    ;COMPARETRUE ds.b #1   ;cleared when compare started,set when true
3    ;--------------------INIT_ADC Subroutine-----------------------
4    INIT_ADC  bsr  ADC_STOP
5              mov  #$00,ADCCFG   ;fast clock settings,8b mode
6              mov  #$20,ADCSC2   ;software trigger, compare < enabled
7              bset ADCPIN,APCTL1 ;enable ADC pin
8              rts
9    ;--------------------COMP_ADC Subroutine-----------------------
10   ;Initiate a capture on PAD pin
11   START_ADC sta  ADCCVL        ;write desired upper compare threshold
12             mov  #FALSE,COMPARE_TRUE ;initialize global variable
13             mov  #(ADCPIN|(1<<AIEN)|(1<<ADCO)),ADCSC1  ;start compare
14             rts
15   ;--------------------ADC ISR-----------------------------------
16   ADCISR    lda  ADCRL         ;clear COCO, ACK interrupt
17             mov  #TRUE,COMPARE_TRUE
18             bsr  ADC_STOP       ;power down ADC
19             rti
```

Code Listing 2.7: Interrupt Driver for 8 Bit ADC Compare Function on Pin PAD0.

The ADC interrupt service routine begins on line 16. The first instruction loads the ADC output from the result register, at the same time clearing COCO as a side-effect, which also acknowledges the interrupt. The MOV instruction on line 17 sets the driver global variable to TRUE to indicate to software that the compare operation was successful, i.e., that the voltage on the selected pin fell below the programmed threshold). At this point, the ISR simply shuts down the ADC and returns. Software can poll the variable COMPARE_TRUE to determine that a compare operation occurred, and process it as required by the application.

2.3.7   8-BIT INTERRUPT-BASED DRIVER WITH HYSTERESIS

One issue that can occur when using the compare function is that noise can cause the compare output to oscillate when the input hovers around the compare value. When using interrupts, as in Code Listing 2.7, this can cause the interrupt service routine to be repeatedly triggered. When polling, this can have an effect similar to switch bounce. One way to overcome this behavior is by introducing hysteresis into the system: when the compare greater than event is detected, we require that the input fall below a lower second threshold before it can trigger another greater than compare event. By setting the second threshold far enough below the first, the oscillating behavior is eliminated. This is the principle used in thermostats and other controllers where it is desirable that the system not be rapidly switched on and off. A simple driver implementing hysteresis is shown in Code Listing 2.8. This driver sets COMPAREVAL to true whenever the voltage on pin PAD0 exceeds a high threshold (HITHRESHOLD); COMPAREVAL is set false when the input falls below LOWTHRESHOLD.

The initialization subroutine configures the basic operation of the ADC in ADCCFG and initializes the driver global variable to false. It initiates a greater-than-HITHRESHOLD capture by calling subroutine CAP_HIGH before returning.

CAP_HIGH and CAP_LOW are similar subroutines that initiate greater-than and less-than captures using the high and low thresholds, respectively. Each subroutine writes the compare threshold value to ADCVL, configures ADCSC2 correct capture type, and initiates a capture with interrupts and continuous conversion enabled.

The interrupt subroutine reads the ADC data register to clear COCO and acknowledge the interrupt. It then tests the ADFGT bit in ADCSC2 to determine if the compare event that triggered the ISR was a greater-than or less-than. If the detected event is less-than, COMPAREVAL is set to FALSE and a CAP_HIGH is called to initiate a greater-than compare before returning; otherwise, COMPAREVAL is set TRUE and CAP_LOW is called. Thus, when the voltage exceeds HITHRESHOLD, the ADC is configured for a less-than capture with LOWTHRESHOLD; likewise, when the voltage falls below LOWTHRESHOLD, the ADC is configured for a greater-than capture with HITHRESHOLD. This implements the desired hysteresis.

2.4   ANALOG COMPARATOR

An analog comparator is a device that compares two analog inputs and produces a Boolean output indicating which is larger. The schematic symbol for an analog comparator is shown in Figure 2.7. The analog inputs are typically referred to as the non-inverting input, V^+, and the inverting input, V^-. The output voltage, V_{OUT}, has the binary transfer function $V_{OUT} = V_{OH}$ (logic high) when $V^+ > V^-$ and $V_{OUT} = V_{OL}$ (logic low) when $V^+ \leq V^-$. Figure 2.7 shows the transfer function for a fixed V^-.

By fixing one of the input voltages to the comparator, a threshold detection circuit is created, as illustrated in Figure 2.8. If the reference voltage V_{REF} is connected to V^+, then the circuit will be a less-than threshold detector (inverting comparator) that outputs high when $V_{IN} < V_{REF}$; connecting V_{REF} to V^- and V_{IN} to V^+ produces a greater than (non-inverting) threshold detector. In Figure 2.8, a voltage divider is used to generate the reference voltage, where $V_{REF} = \frac{R_1}{R_1 + R_2} V_{DD}$. The values of R1 and R2 should be chosen to minimize the power dissipated and should have similar magnitude to the pull-up resistors used in Chapter 4. If V_{REF} needs to be precise, a more precise reference voltage circuit could be used.

In practice, when the input voltage transitions slowly or remains fixed near the reference voltage, noise on the input can cause the digital output to switch rapidly between its high and low values. This oscillating behavior can be overcome by introducing hysteresis into the transfer function of the threshold detector circuit. The basic idea to form a circuit that has two switching points:

```
HITHRESHOLD  equ  $C0    ;75% of VDD
LOWTHRESHOLD equ  $40    ;25% of VDD
; ----------------Required Global Variables-------------------------
;COMPAREVAL ds.b #1  ;cleared when <LOWTHRESHOLD, set when >HITHRESHOLD
;--------------------INIT_ADC Subroutine-------------------------
INIT_ADC     bsr  ADC_STOP      ;stop ADC
             mov  #$00,ADCCFG    ;fast clock settings,8b mode
             bset ADCPIN,APCTL1  ;enable ADC pin
             mov  #FALSE,COMPAREVAL ; initialize global variable
             bsr  CAP_HIGH       ;initialize > compare
             rts
;---------------------CAP_LOW Subroutine-------------------------
;Initiate a falling edge capture with low threshold on PAD pin;
CAP_LOW      mov  #$20,ADCSC2    ;software trigger, compare < enabled
             mov  #LOWTHRESHOLD,ADCCVL
             mov  #(ADCPIN|(1<<AIEN)|(1<<ADCO)),ADCSC1 ;start compare
             rts
;---------------------CAP_HIGH Subroutine-------------------------
;Initiate a rising edge capture with high threshold on PAD pin;
CAP_HIGH     mov  #$30,ADCSC2    ;software trigger, compare > enabled
             mov  #HITHRESHOLD,ADCCVL
             mov  #(ADCPIN|(1<<AIEN)|(1<<ADCO)),ADCSC1 ;start compare
             rts
;---------------------ADC ISR-------------------------------------
ADCISR       lda  ADCRL                 ;clear COCO, ACK interrupt
             brclr 4,ADCSC2,DETECT_LOW ;check if < or > capture
DETECT_HIGH  mov  #TRUE,COMPAREVAL      ;detected > compare
             bsr  CAP_LOW               ;initiate < compare
             bra  ENDADCISR
DETECT_LOW   mov #FALSE,COMPAREVAL      ;detected < compare
             bsr  CAP_HIGH              ;initiate > compare
ENDADCISR    rti
```

Code Listing 2.8: Interrupt Driver for 8 Bit ADC Compare Function with Hysteresis.

V_{HtoL} and V_{LtoH}. For the non-inverting configuration, when the output is high, it will only switch to low when the input falls below V_{HtoL}; when the output is low, it can only switch to high when the input exceeds V_{LtoH}. To prevent oscillations in the output due to noise, it is desirable to have $V_{HtoL} << V_{LtoH}$. The inverting configuration is similar.

The circuits shown in Figure 2.9 illustrate methods for adding positive feedback to the circuits of Figure 2.8, obtaining a threshold voltage that is dependent on V_{OUT}. Note that these circuits are identical to those in Figure 2.8 except for the addition of resistors R_i and R_f. In all circuits, it is assumed for simplicity that $V_{OH} = V_{DD}$ and $V_{OL} = 0$.

For the inverting threshold detector, when $V_{OUT} = V_{DD}$, feedback resistor R_f is in parallel with R_2, lowering the effective resistance of R_2. This can also be thought of as increasing the proportion of R_1 in the voltage divider, thereby increasing V_{REF}. Thus, to switch the output low requires bringing

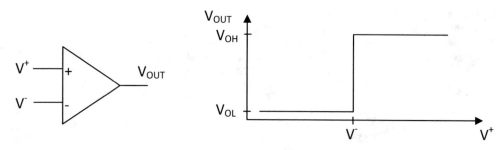

Figure 2.7: Schematic symbol and transfer function of analog comparator.

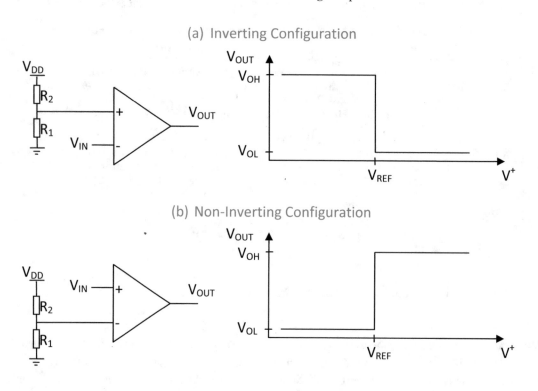

Figure 2.8: Comparator circuits for voltage threshold detection.

the input above the higher $V^+ = V_{REF} = V_{HtoL} = V_{DD}\frac{R_1}{R_1+R_2//R_f}$, where $//$ indicates the parallel combination of the resistance values $\left(R_i//R_j = \frac{R_iR_j}{R_j+R_i}\right)$. On the other hand, when $V_{OUT} = 0$, R_f is in parallel with R_1, decreasing its proportion in the voltage divider and lowering V_{REF}. To switch the output high, the input has to fall below the lower $V^+ = V_{REF} = V_{LtoH} = V_{DD}\frac{R_1//R_f}{R_1//R_f+R_2}$.

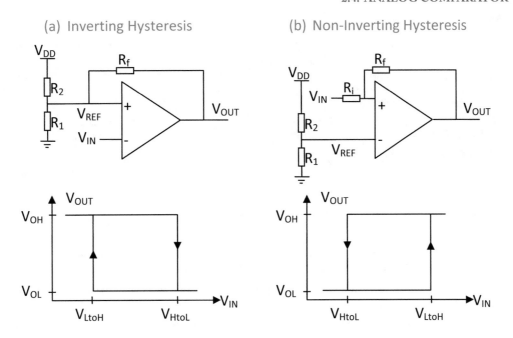

Figure 2.9: Using feedback to achieve hysteresis in threshold detection function.

For the non-inverting threshold detector, V_{REF} is fixed by the voltage divider on the inverting input. The inclusion of R_i and R_f create a voltage divider on V^+. When $V_{OUT} = V_{DD}$, $V^+ = V_{IN} + (V_{DD} - V_{IN}) \frac{R_i}{R_i + R_f}$. To switch the output to low requires that $V^+ \le V_{REF}$, or $V_{IN} + (V_{DD} - V_{IN}) \frac{R_i}{R_i + R_f} \le V_{REF}$. Rearranging, we obtain $V_{in} \le V_{REF} \frac{R_f + R_i}{R_f} - V_{DD} \frac{R_i}{R_f}$, which implies $V_{HtoL} = V_{REF} \frac{R_f + R_i}{R_f} - V_{DD} \frac{R_i}{R_f}$. Following a similar line of reasoning, $V_{OUT} = 0$ creates the voltage divider $V^+ = (V_{IN} - 0) \frac{R_f}{R_i + R_f}$, which leads to $V_{LtoH} = V_{REF} \frac{R_f + R_i}{R_f}$. Clearly, $V_{HtoL} = V_{LtoH} - V_{DD} \frac{R_i}{R_f}$. When solving, we end up with two equations and four unknowns. One method is to let $R_1 = R_2$, which sets $V_{REF} = V_{DD}/2$. Then, the high and low thresholds are equidistant from the power supply rails.

Example 2.6. For the non-inverting comparator circuit, find values for the component resistors to achieve $V_{LtoH} = .8 V_{DD}$.
Solution: Assume $V_{REF} = .5 V_{DD} (R_1 = R_2)$. $V_{LtoH} = V_{REF} \frac{R_f + R_i}{R_i} = .8 V_{DD}$. Thus, $\frac{R_f + R_i}{R_i} = 1.6$. Choosing $R_f = 1$ kΩ gives $R_i = 0.6$ kΩ.
Answer: Choose $R_1 = R_2 = R_f = 1$ kΩ. Then $R_i = 0.6$ kΩ.

Example 2.7. For the inverting comparator circuit, find values for the component resistors to achieve $V_{HtoL} = .8V_{DD}$ and $V_{LtoH} = .2V_{DD}$.

Solution: $V_{HtoL} = V_{DD}\frac{R_1}{R_1+R_2//R_f} = .8V_{DD}$. Thus, `R1=.8(R1+R2//Rf)`.

Rearranging, we get `R1=4(R2//Rf)`. $V_{LtoH} = V_{DD}\frac{R_1//R_f}{R_1//R_f+R_2} = .2V_{DD}$, which gives `R2=4(R1//RF)`. Clearly, `R1=R2`. Substituting, we get `R1=4(R1//Rf)`. This gives `R1=3Rf`.

Answer: Choose `Ri=Rf=1 kOhm`, `R1=R2=3 kOhm`.

2.5   ANALOG COMPARATOR ON THE MC9S08QG4/8

The analog comparator module on the MC9S08QG4/8 microcontroller has the ability to operate with some or all of its inputs and output connected to microcontroller pins. The comparator can be configured to generate an interrupt on the rising edge, falling edge or both edges of the comparator output. The inverting comparator input is connected to pin ACMP$^-$. The non-inverting input can be connected to pin ACMP$^+$, or be internally connected to a bandgap reference voltage that has a nominal value of approximately 1.2V. This bandgap voltage reference is a constant reference that is essentially unaffected by variations in temperature and power supply. The comparator output can be left unconnected, to be used by software only, or routed to pin ACMPO allowing feedback circuits to be connected through the various pins. In the latter configuration, the comparator can operate without software control.

The MC9S08QG4/8 analog comparator module has a single I/O register associated with it: Analog Comparator Status and Control Register (ACMPSC). The format of this register is shown in Figure 2.10. ACME, Analog Comparator Module Enable, is the main control bit that enables the analog comparator. When set, the comparator is enabled and pin ACMP$^-$ is used as the inverting input to the comparator. Since the analog comparator has the highest priority on this pin, the other functions (including PTA2 and PAD2) are disabled. ACBGS, Analog Comparator BandGap Select, selects what is connected to the non-inverting comparator input. When set, the internal bandgap reference voltage (V_{BG}) is selected; when clear, pin ACMP$^+$ is enabled as the non-inverting input to the comparator (disabling the lower priority functions on this pin). ACOPE, the Analog Comparator Output Pin Enable, selects whether the comparator output is connected to pin ACMPO or not. When connected, the output can be used to drive external digital circuits or to incorporate external feedback resistors. When clear, the comparator output is unconnected, but can still be read by software.

ACO, Analog Comparator Output, is a data bit that reflects the binary output of the comparator. This can be read by software to determine whether ACMP$^+$ > ACMP$^-$ (ACO=1) or not. ACF, Analog Comparator Flag, becomes set when the comparator output changes. This can be used to poll for a compare event. ACF is cleared by writing a 1 to the ACF bit. ACMOD selects what type of event will set ACF. When ACMOD is 00 or 10, ACF becomes set on the falling edge of the comparator output;

ACMPSC: Analog Comparator Status and Control ($001A)

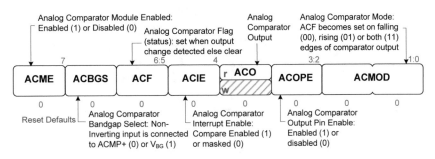

Figure 2.10: Format of the analog comparator status and control register (ACMPSC).

when 01, ACF becomes set on the rising edge; and when 11, ACF becomes set on the rising or falling edge. ACIE (Analog Comparator Interrupt Enable), when set, configures the comparator to issue an interrupt request when ACF=1.

2.6   ANALOG COMPARATOR DRIVER EXAMPLES

Despite its apparent simplicity, the analog comparator on the MC9S08QG4/8 can be used to perform a wide variety of common analog integration tasks. Without software intervention, the comparator module can be used to generate an external rectangular wave from a sinusoid, sawtooth or other analog waveform. With an external resistor and capacitor, it can be used to perform hardware switch debouncing.

2.6.1   DC VOLTAGE MONITORING

Figure 2.11 shows a simple circuit to implement DC voltage monitoring. In this application, the analog comparator is configured to use the internal bandgap reverence voltage $V_{BG} = 1.2V$ for the non-inverting input. Suppose, for example, we are interested in determining when the monitored DC voltage V_{IN} falls below 2.8V. The voltage divider is configured such that when V_{IN} is 2.8, the inverting input will be at 1.2V (equal to the non-inverting input). Below that voltage, the comparator output will become 1 (the inverting input falls below the non-inverting input). The values for R_1 and R_2 were found by noting that $V_{ACMP^-} = V_{IN} \frac{R_1}{R_1 + R_2}$. Setting $V_{ACMP^-} = 1.2V$ and $V_{IN} = 2.8V$ gives $3R_2 = 4R_1$; then selecting $R_1 = 3$ kΩ gives $R_2 = 2$ kΩ.

Code Listing 2.9 shows a skeleton driver for the DC voltage monitoring application. The initialization subroutine simply writes the correct configuration value to ACMPSC to enable interrupts, disable all pins (except ACMP⁻, which is always enabled when ACMP is in use) and configure the module such that ACF becomes set on the falling edge of the comparator output. The skeleton interrupt service routine, which is triggered each time V_{IN} falls below 2.8V (not as long as it is below

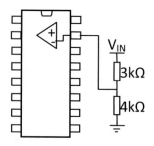

Figure 2.11: MC9S08QG4/8 analog comparator configured as a DC voltage monitor.

```
ACMPSC          equ  $001A           ;location of ACMPSC
ACF             equ  5               ;ACO bit in ACMPSC
;-------------------ACMP Initialization------------------------
;initialize ACMP: module enabled with interrupts, output pin
;                 disabled, ACMP+ connected to internal VBG
;                 falling edge triggered interrupts
INIT_ACMP       mov  #%11010000,ACMPSC
                rts
;-----------------------ACMP ISR-------------------------------
;Skeleton ISR
ACMP_ISR        bset ACF,ACMPSC       ;clear ACF to acknowledge interrupt
                nop                   ;replace with code to perform when
                                      ;voltage falls below 2.8V
                rts
```

Code Listing 2.9: Driver for DC Voltage Monitoring.

2.8V), simply acknowledges the interrupt by clearing the ACF flag in ACMPSC. The NOP instruction would be replaced with the code that would perform the actions necessary upon V_{IN} falling below the threshold.

2.6.2   ANALOG SIGNAL TO DIGITAL WAVEFORM GENERATION

Figure 2.12 shows the MC9S08QG8 analog comparator configured as an analog waveform to digital waveform converter. The comparator is configured as in the non-inverting configuration with hysteresis; the resistor values are those derived in Example 2.6, which provide $V_{HtoL} = .2V_{DD}$ and $V_{LtoH} = .8V_{DD}$. A periodic analog waveform on the input, assumed to vary between 0 to V_{DD}, is converted to a digital output with the same period. The duty cycle of the digital waveform can be determined from the type of analog input waveform and the thresholds.

Code Listing 2.10 shows the driver for the circuit in Figure 2.12. The primary purpose of the driver is to enable and configure the analog comparator module, since the comparator generates this waveform without software assistance. An additional subroutine is provided to allow software

to read the state of the digital waveform output (high or low). The driver initialization subroutine simple writes the configuration value to the ACMPSC register to: enable the ACMP (ACME=1), disable interrupts (ACIE=0), configure the ACMP$^+$ pin as the non-inverting input (ACBGS=0) and enable the output in (ACOPE=1).

```
1    ACMPSC          equ  $001A                  ;location of ACMPSC
2    ACO             equ  3                       ;ACO bit in ACMPSC
3    ;-------------------ACMP Initialization------------------------
4    ;initialize ACMP: module enabled, no interrupts, output pin enabled
5    ;                  ACMP+ pin enabled
6    INIT_ACMP       mov  #%10000100,ACMPSC
7                    rts
8    ;----------------------------GET_ACO----------------------------
9    ;return with current ACMP output in C
10   GET_ACO         brclr ACO,ACMPSC,RTNACO
11   RTNACO          rts
```

Code Listing 2.10: Driver for Digital Waveform Convertor Using Analog Comparator.

The subroutine GET_ACO returns in the C flag the current ACMP output. The BRCLR instruction, since it branches to the next instruction, has no effect other than to place the ACO bit from ACMPSC into C. The subroutine thus returns with C=1 if the current state of the digital output waveform is high, and C=0 otherwise.

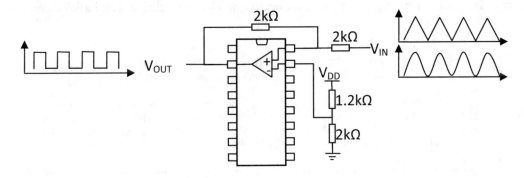

Figure 2.12: MC9S08QG4/8 analog comparator configured as a digital waveform converter.

2.6.3   HARDWARE SWITCH DEBOUNCING USING THE ANALOG COMPARATOR

The analog comparator module can be used to implement the hysteresis needed in the hardware switch debounce circuit of Figure 1.12. One possible circuit configuration is shown in Figure 2.13. In this circuit, the inverting hysteresis configuration was chosen because it requires one fewer resistor.

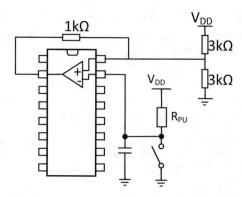

Figure 2.13: Hardware switch debounce circuit using the MC9S08QG4/8 analog comparator.

The resistor values shown are from Example 2.6. The output of the switch output connects to the inverting comparator input ACMP⁻. When the switch is closed, ACMP⁻ = 0V. When opened, the capacitor slowly charges through R$_{PU}$, which filters out switch bounce. When ACMP⁻ reaches the threshold V$_{LtoH}$, the comparator output switches to high and the threshold changes to the lower V$_{HtoL}$. This change in threshold prevents noise from rapidly switching the comparator output.

An interrupt-based driver for the circuit in Figure 2.13 is shown in Code Listing 2.11. The driver is designed to perform an action whenever the switch state changes. The initialization sub-

```
1  ACMPSC          equ  $001A      ;location of ACMPSC
2  ACO             equ  3          ;ACO bit in ACMPSC
3  ACF             equ  5
4  ;-------------------ACMP Initialization------------------
5  ;initialize ACMP: module enabled, interrupts enabled,output pin
6  ;                 enabled, ACMP+ pin enabled, ACF on both edges
7  INIT_ACMP       mov  #%10010111,ACMPSC
8                  rts
9  ;------------------------ADC ISR------------------------
10 ;Skeleton ISR for ACMP Debounced Switch
11 ACMPISR         bset ACF,ACMPSC ;acknowledge interrupt, clear ACF
12                 brset ACO,ACMPSC,CLOSED_SW
13 OPEN_SW         nop             ;replace nop with operations to do
14                 bra ENDACMPISR  ;    when switch opens
15 CLOSED_SW       nop             ;replace nop with operations to do
16                 ;    when switch closes
17 ENDACMPISR      rti
```

Code Listing 2.11: Interrupt-Based Driver for Hardware Debounced using Analog Comparator.

routine configures the analog comparator with interrupts enabled, all pins connected, and with ACF being set on both rising and falling edges of the comparator output.

A skeleton ISR is provided that provides the essential processing for the ACMP interrupt as well as two NOP instructions that can be replaced with whatever code is required to be executed when the switch opens or closes. The interrupt service routine first clears ACF to acknowledge the interrupt request by writing a 1 to the ACF bit in ACMPSC. Then, the analog comparator output is tested through the ACO bit in ACMPSC to determine whether the switch is opened or closed. Because the switch is connected to the inverting input, the comparator output is high when the switch input is low. The switch input is low when closed; thus, ACO=1 indicates that the switch is closed.

2.7   CHAPTER PROBLEMS

1. An 8 bit ADC has V_{RL} tied to 0V and V_{RH} tied to 3V. If the analog input is $V_A = 1.22V$, what is the binary output of the ADC?

2. An 10 bit ADC has V_{RL} tied to 0V and V_{RH} tied to 5V. If the ADC output is \$076, what is

   (a) the minimum value that the analog input voltage, V_A, could be?

   (b) the maximum value that the analog input voltage, V_A, could be?

3. Suppose you have a 3 bit ADC with $V_{RL} = 0V$ and $V_{RH} = 4V$. Complete the following table, where V_A is the input voltage approximation, assuming 000 is 0V and 111 is 7Δ, and V_B is the input voltage approximation assuming 000 is 0V and 111 is V_{RH}

ADC Output	Minimum Input Voltage that could produce this output	Maximum Input voltage that could produce this output	V_A	V_B
000	0 V		0 V	0 V
001				
010				
011				
100				
101				
110				
111		4 V		4 V

4. For the MC9S08QG4/8 ADC module, describe the purpose/meaning of the

   (a) COCO status bit

   (b) ADACT status bit

   (c) Can COCO and ADACT ever be simultaneously 1?

5. For the MC9S08QG4/8 ADC module, list two ways that software can synchronize with the ADC output being ready.

6. Describe the essential difference between the ADC with compare mode enabled and the analog comparator.

   (a) For what type of applications would the ADC compare function be more appropriate?

   (b) For what type of applications would the analog comparator be more appropriate?

7. Find appropriate values for all resistances for the following circuits. Sketch the circuit.

   (a) Comparator in inverting configuration with no hysteresis, $V_{REF} = .45V_{DD}$.

   (b) Comparator in inverting configuration with hysteresis; V_{HtoL} and V_{LtoH} must be between $.5V_{DD}$ and $.6V_{DD}$ apart.

   (c) Comparator in non-inverting configuration with hysteresis; V_{HtoL} and V_{LtoH} must be between $.5V_{DD}$ and $.6V_{DD}$ apart.

8. Sketch two cycles of 3V peak-to-peak sine wave with $+1.5V$ DC offset. On the same sketch, sketch the output of an inverting comparator circuit with hysteresis. Assume $V_{LtoH} = 1V$ and $V_{HtoL} = 2V$.

9. Modify the circuit in Figure 2.11 and the code in Code Listing 2.9 to detect when the voltage V_{IN} exceeds 12V. How far above 12V can V_{IN} exceed before the electrical specifications for the pin exceeded?

CHAPTER 3

Serial Communication

Serial communication is a form of data exchange between digital systems in which multiple bits are transmitted sequentially, one per unit time. Serial interfaces require as few as a single pin to implement, which is their principle advantage over general purpose parallel I/O interfaces. This space advantage, of course, comes at the expense of the increased time it takes to transmit bits sequentially. Most standard serial interfaces used in embedded microcomputers are based on parallel-in serial-out or serial-in parallel-out shift registers that allow the microcontroller to read or write a whole byte at a time, alleviating the need for software to decompose each byte into bits before transmission or recompose bits after reception.

Serial interfaces can be categorized into asynchronous and synchronous types. Synchronous serial communication uses a shared bit clock to mark the presence of a bit on the serial link. Thus, at least two microcontroller pins are required for a synchronous serial interface: one for clock and one for data. In addition, one system, the master, is responsible for generating the common clock used in the communication. The remaining systems, the slaves, operate with respect to this clock. In an asynchronous system, bits are transmitted at a common rate but there is no shared clock to mark the presence of a bit. Such systems use forced signal changes (edges) to allow the communicating endpoints to synchronize their bit clock with the data on the link, which requires the receiver to operate at a higher base clock frequency than the bit clock, allowing the link to be sampled multiple times per bit.

The MC9S08QG8 microcontroller has three serial interfaces: serial communications interface (SCI), synchronous peripheral interface (SPI) and inter-integrated circuit (IIC or I^2C). The SCI is an asynchronous interface, while the SPI and IIC are synchronous interfaces. The MC9S08QG4 has only an I^2C interface, most likely due to the limited package pins available.

3.1 SERIAL COMMUNICATION INTERFACE

The serial communication interface (SCI) is an example of a universal asynchronous receiver/transmitter (UART) circuit. The SCI implements a 1 or 2 wire asynchronous interface between the microcontroller and a peripheral device or microcomputer. Each interface has a receiver (RX) and transmitter (TX); one system's transmitter is connected to the others receiver. Communication between systems can be unidirectional (simplex) or bidirectional (duplex), as shown in Figure 3.1. Full-duplex serial communication allows data to flow in both directions simultaneously, and is achieved via a pair of simplex links in opposite directions. Half-duplex serial communication uses a single link that can operate in one direction at a given time. There is sometimes no coordination as to which direction a half-duplex link is transmitting in at a given time, leading to the possibility that two or more systems attempt to transmit simultaneously. To prevent electrical problems

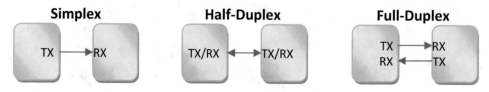

Figure 3.1: Illustration of full-duplex, half-duplex and simplex communication.

caused by having two circuit outputs tied together, such uncoordinated half-duplex serial links are sometimes implemented with open-drain logic (with an external pull-up resistor). Although such a configuration will not lead to electrical problems, bit errors are introduced when two or more systems simultaneously transmit. If open-drain outputs are not used, there must be some coordination between the systems on the link to ensure that only one is transmitting at a given time (as is the case with the MC9S08QG8).

The rate at which data is signaled on a link is called the baud rate, denoted f_{baud}, with units symbols per second. The rate of data transmission is bits per symbol times the baud rate. When binary signaling is used, each symbol conveys one bit and the baud rate is equal to the bit rate. Because there is no common clock, all transmitter and receiver systems on an asynchronous data link must be preconfigured to use the same baud rate. Without this common baud rate assumption, there is no way for such a system to delineate individual bits.

Often, an asynchronous serial link is required to span physically separate systems. In these cases, specific electrical conditions are required for reliable data transmission from one system to another. Standards have been established for such communications. One example is the well-known Electronic Industries Association (EIA) Recommended Standard 232 (RS-232), which defines specifications for the electrical, mechanical and protocol interconnections between systems. RS-232 signals, for example, use voltage-level signaling with the range +3V to +15V indicating a logic 1 and −3V to −15V indicating a logic 0. Digital systems typically use ground and V_{DD} as input/output logic levels. These systems require special interface circuits to convert the standard logic levels to RS-232 signal levels. Within a common circuit, standard logic level signaling can be used.

Asynchronous serial communication proceeds in frames of bits, each frame representing a unit (typically one byte) of data. One frame of a typical asynchronous serial transmission is shown in Figure 3.2, showing 1 start bit, 1 stop bit, 8 data bits and an even parity bit. The asynchronous link has one logic state designated as the idle state, which the link remains in when no data is being transmitted. When open-drain drivers are used, the idle state is typically high. A frame transmission begins with a start bit, which has the opposite logic level as the idle state and lasts for one bit period (inverse of baud rate). The receiver(s) use the edge at the beginning of the start bit to mark the initiation of a frame. Because the baud rate is known, the receiver can use the location of the start bit's falling edge to synchronize its baud clock with the data stream. Each bit period has time $t_{bit} = 1/f_{baud}$. Ideally, the receiver would sample the line multiple times per bit and use an

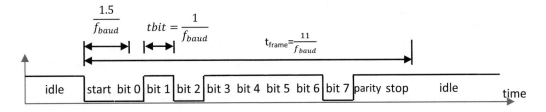

Figure 3.2: Oscilloscope view of asynchronous serial transmission of the byte $7A with 1 start bit, even parity and 1 stop bit.

averaging or voting scheme to eliminate noise. In the simplest case, each bit is sampled once, ideally in the middle of the bit period. The center of data bit zero (the least significant bit) starts 1.5 bit periods after the start bit's falling edge. The center of each subsequent bit occurs 1 bit period after the previous center. Slight variations in the baud clocks of the receiver and transmitter can be tolerated because the receiver resynchronizes at the beginning of each data frame.

Following the start bit, multiple data bits are transmitted in least-significant-first bit order. Typically, 7 or 8 data bits are transmitted per frame. Following the data bits is an optional error checking bit, called the parity bit. The parity bit is usually computed using an exclusive-or of the data bits (even parity) or exclusive-nor (odd parity). In addition, space and mark parity bits are sometimes used. The parity names *even* and *odd* refer to the total number of 1 bits in the frame, counting data bits only. Exclusive-or parity will result in 1 if the total number of data bits is odd and 0 if even; thus, the total number of ones in data and parity bits will always be even. Space parity always sets the parity bit to 0 and mark parity to 1. The receiver can compute the parity and use the transmitted parity bit as a check to see if a bit error has occurred (due to noise, simultaneous transmissions on a half-duplex link or some other problem). Note that parity can detect only certain types of error patterns, and some errors go undetected. Finally, the frame transmission ends with the line going idle for one or more bit periods. These bits are referred to as stop bits and serve two purposes. The first is to mark the end of the frame; the presence of this end-of-frame mark can be checked at the receiver to verify that the end of frame has been reached and that no framing errors have occurred. The second reason for having at least one stop bit is to force the line to return to its idle state to allow the next start bit (edge) to be detected and the receiver resynchronize.

3.1.1   MC9S08QG8 SCI

The SCI on the MC9S08QG8 can operate in simplex, full-duplex or half-duplex mode, has a programmable baud rate and configurable frame format (parity type, number of data bits, number of stop bits). There are two SCI pins that allow separate transmit data (TxD) and receive data (RxD) for full duplex communication. Interrupt synchronization or polling can be used to detect when data has been received or when a data transmission has completed. Interrupts can be separately enabled for receive and transmit functions as well as when an error or idle line condition is detected. Thus, there are three SCI interrupt vectors.

The first step in configuring the SCI module is to set the baud rate, which is accomplished via the SCI baud rate control registers (shown in Figure 3.3). Together, the 13 bit value in these

SCIBDH: SCI Baud Rate Control High Byte (memory mapped at address $0020)

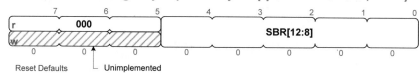

SCIBDL: SCI Baud Rate Control Low Byte (memory mapped at address $0021)

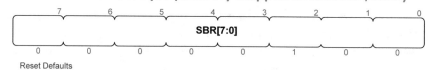

Figure 3.3: Format of the SCI baud rate control registers.

registers, called BR, controls the baud clock divisor. When BR is 0, the baud clock is disabled. For any other value of BR, the relationship between the bus clock and baud rate is $f_{baud} = \frac{f_{bus}}{16 \times BR}$.

Example 3.1. Compute the baud rate divisor value required to configure the SCI to operate at 9600 baud, assuming the bus clock is 4 MHz. Write the instruction(s) necessary to configure the baud clock.

Solution: $f_{baud} = \frac{f_{bus}}{16 \times BR}$; therefore $BR = \text{round}\left(\frac{f_{bus}}{16 \times f_{baud}}\right)$. Substituting 4 MHz and 9600 baud gives BR=26.

Answer ldhx #26 ;baud rate divisor for 9600 Bd
        sthx SCIBDH ;configure baud rate

The second step needed to configure the SCI is to establish the basic communication parameters, such as duplexity and frame format. Figure 3.4 shows the format of the three SCI configuration registers that control these parameters. Five bits from these registers, LOOPS (Loop Mode Select), RSRC (Receiver Source Select), TE (Transmitter Enable), RE (Receiver Enable) and TXDIR (TxD pin Direction) control the duplexity of the channel. Table 3.1 summarizes the 7 possible operating modes of the SCI that result from configuration of these bits. The first column, LOOPS, determines whether the transmitter and receiver are internally tied together (looped back) or separately brought out to the microcontroller pins. When LOOPS=0, the transmitter and receiver are separately brought

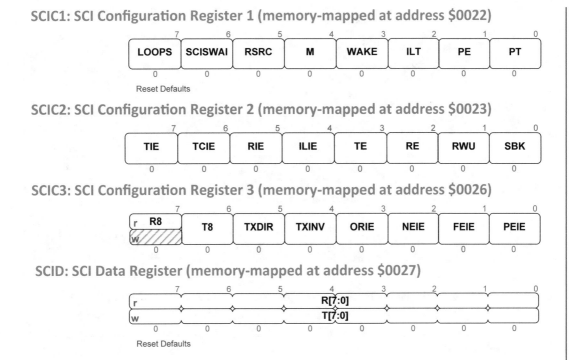

Figure 3.4: Format of SCI configuration and data registers.

out to microcontroller pins TxD and RxD, respectively. If both the transmitter and receiver are active, a full-duplex channel results; otherwise either an output or input simplex channel results. In loopback mode (LOOPS=1), the SCI receiver is disconnected from the microcontroller pin and internally connected to the transmitter. Loopback mode is used to achieve half-duplex operation, with the direction of the channel controlled by TXDIR. Note that in half-duplex mode, transmitted data is also internally looped back, so the data will appear in the receive register.

The SCI Data register, SCID, provides an interface to the SCI transmit and receive buffers. A read from SCID loads the value currently in the SCI receive buffer; a store writes the value to the SCI transmit buffer. Both the receiver and transmitter are double buffered. This means that the SCI can be receiving a new byte while the last received byte is waiting to be read by software, or that software can write another byte to be queued for transmission while a previous value is being transmitted. From the software perspective, it takes two writes to an idle SCI before the transmit data register is marked full. This also means that software has almost one full serial frame time to read a byte after receiving it without the possibility of it being overwritten by newly received data.

The transmitted frame format is configured via bits M (8 or 9 bit Mode select), PE (Parity Enable), PT (Parity Type) and T8 (transmit data bit 8). Note that an MC9S08QG8 SCI frame

Table 3.1: SCI channel direction configuration options.

LOOPS	RSRC	TE	RE	TXDIR	Mode
0	-	0	0	-	SCI not enabled.
0	-	0	1	-	Simplex (SCI receive only). RxD pin enabled, TxD pin available for GPIO.
0	-	1	0	-	Simplex (SCI receive only). TxD pin enabled, RxD pin available for GPIO.
0	-	1	1	-	Full-duplex mode. Both RxD and TxD pin functions enabled.
1	0	-	-	-	Internal loopback mode. RxD and TxD pins not connected and available for GPIO.
1	1	-	-	0	Half-duplex mode using TxD pin only; channel is input. RxD pin not used (available as GPIO).
1	1	-	-	1	Half-duplex mode using TxD pin only; channel is output. RxD pin not used (available as GPIO).

always has 1 start bit; this parameter is not configurable. The M bit selects whether 8 or 9 bit data is to be transmitted (including the optional parity bit); when M=1, and parity is not being used, the ninth data bit transmitted is placed in the T8 bit of SCISC3 (the ninth data bit received is read from the R8 bit of SCIC3). In 9 bit modes, the ninth data bit can be used for either transmission of an extra bit of data or can be used to implement either mark or space parity or an extra stop bit (when parity is disabled). Table 3.2 lists the configurable SCI frame formats as well as the location of the data.

Table 3.2: SCI frame formats.

M	PE	PT	T8	Start Bits	Data Bits	Parity Bits	Stop Bits
0	0	-	-	1	8 (SCID[7:0])	0	1
0	1	0	-	1	7 (SCID[6:0])	1 (Even)	1
0	1	1	-	1	7 (SCID[6:0])	1 (Odd)	1
1	1	0	-	1	8 (SCID[7:0])	1 (Even)	1
1	1	1	-	1	8 (SCID[7:0])	1 (Odd)	1
1	0	-	0	1	8 (SCID[7:0])	1 (Space)	1
1	0	-	1	1	8 (SCID[7:0])	1 (Mark)	1
1	0	-	1	1	8 (SCID[7:0])	0	2
1	0	-	msb	1	9 (T8 + SCID[7:0])	0	1

The SCI has eight status flags contained in the SCI Status Register. The format of this register is shown in Figure 3.5. Each of these flags can be independently configured to generate an interrupt request. The SCI has three interrupt vectors: SCI transmit, SCI receive and SCI error. The flags,

SCIS1: SCI Status Register 1 (memory-mapped at address $0024)

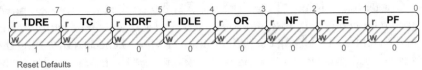

Reset Defaults

Figure 3.5: Format of the SCI status register.

their interrupt enables, associated interrupt vectors and method to acknowledge the interrupt (by clearing the status flag) are summarized in Table 3.3. TDRE (transmit data register empty) becomes

Table 3.3: SCI status flags and associated interrupt vectors.

Flag	Interrupt Enable	Interrupt Vector	Steps to Acknowledge IRQ (Clear the Flag)
TDRE	TIE (SCIC2)	SCI Transmit	Read SCIS1 followed by write SCID
TC	TCIE (SCIC2)	SCI Transmit	Read SCIS1 followed by write SCID
RDRF	RIE (SCIC2)	SCI Receive	Read SCIS1 followed by read SCID
IDLE	ILIE(SCIC2)	SCI Receive	Read SCIS1 followed by read SCID
OR (overrun error)	ORIE (SCIC3)	SCI Error	Read SCIS1 followed by read SCID
NE (noise error)	NEIE (SCIC3)	SCI Error	Read SCIS1 followed by read SCID
FE (framing error)	FEIE (SCIC3)	SCI Error	Read SCIS1 followed by read SCID
PE (parity error)	PEIE (SCIC3)	SCI Error	Read SCIS1 followed by read SCID

set when it is safe to write a new value to the SCI data register SCID. The SCI circuitry contains two buffers. The transmit data buffer is written by software (via write to SCID) and holds the next byte to be transmitted. There is a second internal shift register that holds the byte currently being transmitted. Because of this, it can initially take two writes to SCID before TDRE clears (indicating another value cannot be written). TC (transmission complete) becomes set at after an entire frame has been transmitted (including the parity and stop bits). Both TDRE and TC are cleared by writing a new data value to SCID after a read from SCIS1. An SCI transmit interrupt is requested whenever TDRE=1 or TC=1, provided that the corresponding enable bit in SCIC2, TIE or TCIE, is set.

RDRF (receive data register full) becomes set when a character has been received and is waiting to be read from the SCI data register. The IDLE flag becomes set when an idle condition is detected on the line, defined as the line being idle for a full frame period. Similar to TDRE and TC, RDRF and IDLE are cleared by reading SCIS1 followed by an access (load) from SCID. A receive interrupt is enabled by setting the RIE or ILIE bits. The receive interrupt request is generated whenever RIE=1 and RDRF=1, or whenever ILIE=1 and IDLE=1.

The remaining four flags are error flags. The noise flag becomes set when noise is detected on the receiver, even though the noise might not have resulted in a transmission error. The Framing Error (FE) flag becomes set when an error is detected in the stop bit. This could be a result of noise, mismatched baud rates, hardware error, or simply the receiver circuitry being enabled in the middle of a character transmission. The Parity Error (PE) flag is raised whenever the received parity bit does not match the computed parity bit. Note that when Mark and Space parity are implemented via the ninth data bit (see Table 3.2), the parity must be manually checked by testing the R8 bit in SCIC3. Finally, an overrun error occurs when new data is received in the SCI and is ready to be transmitted to the receive data register, but the receive data register is not empty. In this case, the overrun error (OR) flag becomes set and the newly arrived data is discarded. The SCI error interrupt services the four error flags (overrun, noise, framing and parity). Each of these flags can be configured to generate an SCI Error interrupt request. There are four corresponding enable bits (ORIE, NEIE, FEIE and PEIE, respectively) that control which of these flags, when set, generates an interrupt request.

The remaining configuration bits control miscellaneous and less common options. SCISWAI (bit 6 of SCISC1) controls whether the SCI is deactivated while the CPU is in wait mode; setting this bit stops the SCI clocks in wait mode. TXINV, when set, inverts the polarity of the transmitted and received frames (idle and stop become 0, start becomes 1 and all data and parity bits are inverted). Refer to the microcontroller datasheet for additional information.

3.1.2 SCI DRIVER EXAMPLES

As evidenced by the large number of configuration registers and three dedicated interrupt vectors, the SCI is a highly configurable device that allows great flexibility of use. It would not be possible to cover the wide range of possible driver scenarios in this section. Instead, three drivers are presented from which other configurations can be derived. The first driver is a simple full-duplex polled-I/O driver with transmit and receive functions. The second driver is an interrupt-based driver that illustrates the use of a ring-buffer to hold received data. The final driver example applies the ring-buffer approach to an interrupt-based transmit driver, which is slightly more complex. All of the drivers presented use full-duplex or simplex communications. The HCS08 SCI does not operate with open-drain outputs when in half-duplex mode (an external circuit must be used to implement this). For a half-duplex channel, either a protocol must be established to ensure that both sides of the channel do not transmit simultaneously or hardware mechanisms must be in place to ensure that if this happens there will be no harm to the circuitry, i.e., an open-drain channel or a form of short-circuit protection). In many cases, the interface from the microcontroller to the serial line driver circuitry will be full-duplex or simplex even if the channel itself is operated in half-duplex mode.

The SCI driver common memory map definitions are shown in Code Listing 3.1. In addition to these definitions, interrupt-based drivers require that the appropriate interrupt vectors be initialized. Lines 1 through 11 define the SCI I/O register locations as well as the location of status and control bits used by the driver examples in this section. Lines 12 through 14 define the baud rate that the SCI will use. Line 12 is where the desired baud rate is defined. Since the baud rate is derived from the bus clock, the bus clock frequency must be accurately defined on Line 13. The assembler expression (FBUS/FBAUD/16) on line 14 uses the desired baud rate and defined bus clock to compute the baud

```
 1    SCIBDH      equ    $0020      ;memory mapped SCI I/O registers
 2    SCIBDL      equ    $0021
 3    SCIC1       equ    $0022
 4    SCIC2       equ    $0023
 5    SCIS1       equ    $0024
 6    SCIC3       equ    $0026
 7    SCID        equ    $0027
 8    TDRE        equ    7          ;transmit data register empty SCIS1
 9    TC          equ    6          ;transmit complete SCIS1
10    RDRF        equ    5          ;receive data register full SCIS1
11    TIE         equ    7          ;transmit interrupt enable SCIC1
12    FBAUD       equ    9600       ;desired baud rate for SCI (Bd)
13    FBUS        equ    4000000    ;trimmed clock frequency (Hz)
14    BAUDDVSR    equ    (FBUS/FBAUD/16);baud rate divisor value
```

Code Listing 3.1: Common Memory Map Definitions for SCI Driver Examples.

rate divisor; this divisor is to be written into the SCIBDH:SCIBDL registers to configure the desired baud rate.

3.1.3 POLLED-I/O FULL-DUPLEX SCI DRIVER

Code Listing 3.2 is an example of the simplest type of SCI driver: a polled-I/O driver on a full-duplex channel. That the channel is full-duplex implies that it is both possible and safe for the SCI to transmit and receive data simultaneously. This means that after writing a byte for transmission, the software does not need to block until the transmission is complete. This driver illustrates basic transmit and receive operations and error checking.

Driver subroutine INIT_SCI configures the SCI to operate in full-duplex mode (both RX and TX enabled) with the desired baud rate and with a frame format the uses 8 data bits, an even parity bit and one stop bit. Of course, the device at the other end of the channel must use the same frame format and baud rate.

The first driver interface subroutine, WAITBYTE, is a subroutine that blocks until a byte is received by the SCI. If a received byte is already waiting in the receive data register it will be returned immediately; otherwise, the subroutine will block until a byte is received. Lines 16 and 17 represent the polling loop that waits until RDRF=1, indicating that a byte has been received and is waiting to be read via the SCID register. The watchdog timer reset is necessary because the polling loop can run indefinitely. Once the poling loop is exited, the load of the byte from SCID also clears the RDRF flag (recall that clearing the flag requires a read from RDRF followed by a read from SCID, and polling RDRF serves as the necessary read). The subroutine returns with the received byte in A.

The second driver subroutine, GETBYTE, performs a non-blocking receive operation with checks for parity and framing errors. The subroutine returns the byte received in A with C=0 if the byte was received error free; the return value C=1 indicates that either no byte was received or a byte was received but an error was detected (the value in A should not be used). The BRCLR instruction on

```
1   INIT_SCI      pshh                ;callee save
2                 pshx
3                 ldhx  #BAUDDVSR     ;load baud rate divisor
4                 sthx SCIBDH         ;write baud rate divisor
5                 mov #$12,SCIC1      ;even parity, 8b, no loopback
6                 mov #$0C,SCIC2      ;no interrupts, enable TxD,RxD
7                 mov #$00,SCIC3      ;no interrupts, normal polarity
8                 pulx
9                 pulh                ;callee restore
10                rts
11  ;------------------------------------------------------------
12  ; Subroutine Waitbyte()
13  ;      blocks until byte received by SCI
14  ;      returns byte value in A (does not check for errors)
15  ;------------------------------------------------------------
16  WAITBYTE      sta WATCHDOG        ;reset watchdog in polling loop
17                brclr RDRF,SCIS1,WAITBYTE ;poll until RDRF=1
18                lda  SCID           ;load received byte, clear RDRF
19                rts
20  ;------------------------------------------------------------
21  ; Subroutine Getbyte()
22  ;      non-blocking receive: value received is returned in A
23  ;      returns C=0 if value received error free; C=1 otherwise
24  ;------------------------------------------------------------
25  GETBYTE       brclr RDRF,SCIS1,GETBRTN ;RDR empty
26                lda  SCIS           ;load status register
27                and  #$03           ;test
28                bne  GETBERR        ;error if PE or FE
29                clc                 ;return C=0 no error
30                bra  GETBRTN
31  GETBERR       sec                 ;C=1 error or no byte received
32  GETBRTN       lda  SCID           ;load received byte, clear flags
33                rts
34  ;------------------------------------------------------------
35  ; Subroutine PUTBYTE()
36  ;      blocks until SCI can accept byte to transmit
37  ;      then writes value in A to SCI data register and returns
38  ;------------------------------------------------------------
39  PUTBYTE       sta WATCHDOG        ;reset watchdog in polling loop
40                brclr TDRE,SCIS1,PUTBYTE ;poll until TDRE=1
41                sta SCID            ;write tx data, clear TDRE
42                rts
```

Code Listing 3.2: Full-Duplex Polled SCI Driver (8B, Even Parity).

Line 25 branches to the error return block if no byte has been received. Lines 26-28 check the parity and framing error flags and branch to the error return block if either is non-zero. SCID is loaded into A on line 32 before returning, whether or not a valid data value was received. Performing this load is necessary in all cases to clear the SCI receive and error flags to reset them for the next byte received.

One transmit subroutine, PUTBYTE, waits until the transmit data register is empty (TDRE=1), then writes a new value to the SCID register to queue it for transfer. Although the subroutine waits for TDRE=1, it is not blocking since it does not wait for the value to be transmitted. In fact, the write to TDRE only queues the data for transmission since the transmitter is double buffered. Once TDRE=1, the data value passed in A is written to SCID to queue it for transmission (and to clear TDRE) before the subroutine returns.

3.1.4   INTERRUPT-BASED RING-BUFFERED SCI SIMPLEX (RECEIVE-ONLY) DRIVER

Although the SCI transmitter and receiver are double buffered, there are two scenarios when it might still be necessary for additional bytes of data to be buffered. Consider a system that is using the driver in Code Listing 3.2. In the first scenario, it is possible for more than one byte to be received before software has a chance to call GETBYTE or WAITBYTE. In this case, the program is not operating fast enough to keep up with I/O and an overrun error will occur. The result is that the newly arrived byte(s) are discarded. If more buffering were available, overrun would not occur until the buffer fills up. The other possibility occurs when data is being transmitted by the program and the program calls PUTBYTE more than two time in succession, faster than the data can be transmitted. Although no data is lost in this case, the program is stalled each time PUTBYTE is called waiting for the transmit data register to empty. If more buffering were available, the program would be able to continue unimpeded at least until the extra buffering is exhausted.

Note that buffering can only help when there is a temporary mismatch between I/O and program timing requirements. If data continually arrives faster than a program can process it or is produced faster than it can be transmitted, any size buffer will eventually fill up.

To add buffering to the SCI transmit or receive operations requires interrupt synchronization. On the receive side, newly arrived bytes need to be buffered soon after they are received to prevent overrun. On the transmit side, as soon as a byte completes transmission another byte from the buffer should begin transmission. In other words, the buffering operations need to be performed outside of the control of the main program.

A first-in-first-out (FIFO) queuing policy is required for transmit and receive buffers to maintain data ordering. A linear buffer is not well suited for a FIFO queue because the buffer needs to be compacted each time data is removed from the head of the queue. Linked lists, another form of implementation, result in too much overhead in an embedded system. The simplest form of FIFO queue to implement is the circular (or ring) buffer, which is illustrated in Figure 3.6. The circular buffer is formed by maintaining a head and tail pointer into a linear buffer (representing the head and tail of the queue). When the buffer has a constant size, the head and tail can be maintained

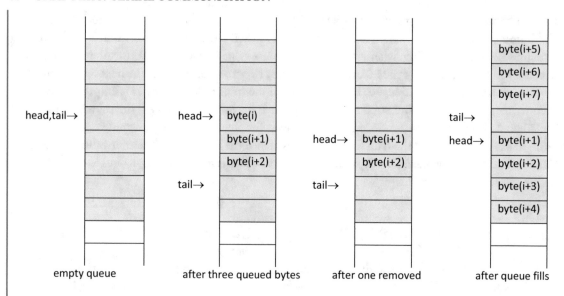

Figure 3.6: Memory map of 8-byte circular buffer in various states of use.

as indexes into the buffer rather than pointers. The head always points to the oldest queued data; the tail always points one index ahead of the last byte queued i.e., the next "empty" buffer space). An empty queue is indicated by the head and tail pointing to the same location. The head and tail are always advanced by one location; incrementing the head and tail is done modulo the size of the buffer so that they wrap around the end of the buffer to the beginning (thus providing the circular characteristic).

Adding a new item to a non-full queue consists of writing the data to the location indicated by the tail, then incrementing the tail. When incrementing the tail would make it equal to the head, the queue is full and cannot be written to. Having the tail always point to an empty space, even when the queue is full, allows the insertion algorithm to write the data to the tail position first, then determine if the queue is full; if full, the tail is not incremented and the write has no effect. Removing and item from a non-empty queue consists of reading the item at the head and advancing the head. Note that this policy results in one queue space always being empty. Although methods exist to allow the entire buffer to be used, the overhead can wastes more memory than is recovered, especially for a queue of bytes.

Code Listing 3.3 provides an example of an interrupt-based SCI receiver using a circular queue. To keep the size of the code listing down, a transmit function was not included; the driver could easily be extended to handle a buffered or unbuffered transmitter using one of the other SCI examples in this section. The interrupt service routine, triggered upon receipt of a byte from the SCI, adds the byte at the tail of the queue (unless it is full). A driver interface subroutine, GETBYTE,

```
1    ;-----------------------------------------------------------
2    ;Required (Direct Page) Global Variables
3    ;RXBUF      DS.B    BUFLEN ;receive buffer BUFLEN-1 bytes
4    ;RXHEAD     DS.B    1      ;head index of buffer (0 to BUFLEN-1)
5    ;RXTAIL     DS.B    1      ;tail index of buffer (0 to BUFLEN-1)
6    ;RXOR       DS.B    1      ;global overrun flag for queue
7    ;-----------------------------------------------------------
8    INIT_SCI    pshh
9                pshx
10               ldhx #BAUDDVSR        ;baud rate divisor
11               sthx SCIBDH           ;write baud rate divisor
12               mov #$12,SCIC1        ;even parity, 8b, no loopback
13               mov #$24,SCIC2        ;Rx and RIE enabled, no TX
14               mov #$00,SCIC3        ;no SCI Error interrupts
15               clr RXHEAD            ;clear ring buffer head
16               clr RXTAIL            ;and tail
17               clr RXOR              ;clear overrun flag
18               pulx
19               pulh
20               rts
21   ;-----------------------------------------------------------
22   ; Subroutine Getbyte()
23   ;      if RXBUF not empty,returns C=0 and next buffered byte in A
24   ;      else returns C=1
25   ;-----------------------------------------------------------
26   GETBYTE     pshx
27               pshh
28               sei                   ;mask interrupts
29               clrh
30               ldx    RXHEAD         ;load buffer head position
31               cpx    RXTAIL         ;compare to tail position
32               beq    BUFEMPTY       ;if equal then rxbuf empty
33               lda    RXBUF,x        ;else get byte at head
```

Code Listing 3.3: Interrupt-Based Ring-Buffered SCI Simplex (Receive) Driver (*Continues*).

returns a byte from the queue if one is available. Four global variables are required to be declared (as shown on lines 3-6); except for RXOR these variables should not be accessed directly outside of the driver. RXBUF is the space used for the circular buffer. BUFLEN should be defined with an equ pseudo-operation to define the buffer size. A total of BUFLEN-1 bytes can be queued in the buffer; in addition, there can be one byte in the SCI receive data register and another in the SCI shift register (being received). Thus, at least BUFLEN+1 bytes must arrive before the queue can overrun. This should help the engineer determine an appropriate size for BUFLEN. RXHEAD and RXTAIL are the index of the buffer head and tail locations, respectively. RXOR is a global overrun flag that can be tested by software to determine if the transmit queue has overrun. Software is responsible for clearing RXOR after it is tested.

```
34              bsr     INCMODN         ;advance head
35              stx     RXHEAD          ;save it
36              clc                     ;set return C=0 (data in A)
37              bra     ENDGETB
38   BUFEMPTY   sec                     ;set return C=1 (empty)
39   ENDGETB    cli                     ;reenable interrupts
40              pulh
41              pulx
42              rts
43   ;------------------------------------------------------------
44   ; SCI RX ISR
45   ;     if RXBUF not full and no parity or framing errors, SCID is
46   ;     added to tail of ring buffer
47   ;------------------------------------------------------------
48   SCIRXISR   pshh                    ;h not auto stacked
49              clrh
50              lda     SCID            ;load received data
51              ldx     RXTAIL          ;load buffer tail position
52              sta     RXBUF,X         ;optimistically store at ring tail
53              bsr     INCMODN         ;increment tail (modulo BUFLEN)
54              cpx     RXHEAD          ;compare new tail to head
55              beq     RXFULL          ;if equal ring is full
56              lda     SCIS1
57              bit     #$03            ;check framing and parity error
58              bne     RTISCIRX        ;if set then error occurred
59              stx     RXTAIL          ;else update tail to save write
60              bra     RTISCIRX
61   RXFULL     mov     #TRUE,RXOR      ;set global overrun flag true
62   RTISCIRX   lda     SCIS1           ;load SCIS1 followed by SCID to
63              lda     SCID            ;   clear RDRF and ERROR flags
64              pulh
65              rti
66   ;------------------------------------------------------------
67   ; INCMODN(X)
68   ;     Increments X modulo BUFLEN  (i.e. X=(X+1) mod BUDLEN)
69   ;------------------------------------------------------------
70   INCMODN    incx                    ;increment x
71              cpx     #BUFLEN
72              bne     ENDINCMODN      ;if not equal to BUFLEN do nothing
73              clrx                    ;else x=0
74   ENDINCMODN rts
```

Code Listing 3.3: (*Continued*) Interrupt-Based Ring-Buffered SCI Simplex (Receive) Driver.

Driver subroutine INIT_SCI configures the operating mode of the SCI and initializes driver variables. RXHEAD and RXTAIL are set equal to indicate an initially empty buffer. RXOR is cleared.

GETBYTE is the interface subroutine used by the program to load a received byte from the buffer. GETBYTE is non-blocking and uses the C flag to indicate whether or not a byte is returned in A. Interrupts are masked during execution of the subroutine, to ensure atomic access to the circular queue in the case where a received byte triggers the ISR during execution of GETBYTE. After callee saving and disabling interrupts, RXHEAD and RXTAIL are compared (lines 30 to 32) to see if the buffer is empty. If so, the subroutine returns with the carry flag set. Otherwise, the head byte is loaded into A (line 33) and removed from the queue on lines 34 to 35. Removing the head byte is accomplished by calling the driver subroutine INCMODN, which increments the value in index register X (modulo BUFSIZE). The value is then stored in RXHEAD.

The interrupt service routine is triggered by RDRF=1, which indicates that a byte is waiting to be read in the SCI receive buffer. The interrupt service routine first loads the received SCI data (line 50) and stores it to the tail position in the queue (line 52); this does not, however, add the byte to the tail. This speculative store assumes there is room in the queue for the new byte. Until the tail pointer is updated the byte has not been added to the queue. The tail will only subsequently be updated if the queue is not full and if there were no SCI errors detected. Lines 53-55 check to see if the queue is full and branch to the end of the subroutine if so, without updating the tail. If not, then the error flags are checked (lines 56-58) and if PE or FE are set the code branches to the end of the subroutine, again without updating the tail. Otherwise, the queue is not full and there were no receive errors so the queue tail index RXTAIL is updated on line 59. The return block, beginning on line 60, loads SCIS1 followed by SCID to clear the SCI receive flags.

Driver subroutine INCMODN increments the head or tail pointer contained in X. After the increment, an IF-statement is used to reset X to 0 if it is equal to BUFLEN, implementing the modulo-BUFLEN function.

3.1.5 INTERRUPT-BASED RING-BUFFERED SCI SIMPLEX (TRANSMIT) DRIVER

Code Listing 3.4 provides an interrupt-based driver for the SCI transmitter using a circular queue. In this case, however, driver interface subroutine PUTBYTE adds a byte to the tail of the queue while the interrupt service routine, triggered by the flag condition TDRE=1, removes a byte from the head of the queue if the queue is not empty. This introduces a slight difficulty when the queue becomes empty or when an empty queue needs to be filled. Consider the case where the queue is initially empty. TDRE=1 (since there is no data being transmitted) and the ISR would be triggered. Since the queue is empty, the ISR cannot initiate a transmission, so it must exit with TDRE still being set (starting a transmission clears TDRE). The ISR would continually re-execute and the main program would make no progress. Thus, the transmit interrupt enable (TIE) bit in SCISC1 must be clear to prevent SCI transmit interrupts until the queue is non-empty. When the queue starts to fill, TIE

```
 1   ;------------------------------------------------------------
 2   ;Required Global Variables
 3   ;TXBUF      DS.B   BUFLEN ;receive buffer length bytes
 4   ;TXHEAD     DS.B   1       ;head index of buffer (0 to BUFLEN-1)
 5   ;TXTAIL     DS.B   1       ;tail index of buffer (0 to BUFLEN-1)
 6   ;------------------------------------------------------------
 7   INIT_SCI   pshh
 8              pshx
 9              ldhx #BAUDDVSR        ;baud rate divisor
10              sthx SCIBDH           ;write baud rate divisor
11              mov #$12,SCIC1        ;even parity, 8b, no loopback
12              mov #$08,SCIC2        ;Tx enabled, Rx disabled
13              mov #$00,SCIC3        ;no error interrupts
14              clr TXHEAD            ;clear ring buffer head
15              clr TXTAIL            ;and tail
16              pulx
17              pulh
18              rts
19   ;------------------------------------------------------------
20   ; Subroutine Putbyte()
21   ;     if TXBUF not full,returns C=0 and transmits or queues byte
22   ;     in A; else returns C=1
23   ;------------------------------------------------------------
24   PUTBYTE    pshx
25              pshh
26              sei                   ;mask interrupts
27              ldx    TXTAIL
28              cpx    TXHEAD
29              bne    QUEUEIT        ;if ring not empty, queue byte
30              brclr TDRE,SCIS1,QUEUEIT ;if TX busy, queue byte
31              sta    SCID           ;else write to SCI data register
```

Code Listing 3.4: Interrupt-Based Ring-Buffered SCI Simplex (Transmit) Driver (*Continues*).

must be set to ensure that each successive byte transmission triggers the ISR to remove another byte from the queue. Also, when the queue empties once again, TIE must be again cleared.

The driver initialization subroutine, INIT_SCI, is similar to the initialization subroutine for the interrupt-based receiver driver. Driver subroutine PUTBYTE accepts a value in A and either queues the byte if the queue is not empty; transmits the byte if the queue is empty and the transmit data register is empty (TDRE=1); or queues the byte if the queue is empty but the transmit data register is full (TDRE=0). If the queue is full, the subroutine returns with C=1 to indicate that the byte was not queued. Since this can be determined after each call to PUTBYTE, there is no global variable to indicate overrun as in the previous code listing.

On lines 26 through 28, the queue head and tail, RXHEAD and RXTAIL, are checked to see if the queue is non-empty. If so, the subroutine proceeds at the label QUEUEIT. Otherwise, since the queue is empty, TDRE is checked on line 30 to determine if the byte can be transmitted. If it cannot

```
32              clc                     ;clear C to indicate success
33              bra     ENDPUTB         ;and goto finish
34  QUEUEIT     bset    TIE,SCIC2       ;enable TDRE interrupts
35              clrh
36              ldx     TXTAIL          ;load buffer tail
37              sta     TXBUF,X         ;optimistically store at ring tail
38              bsr     INCMODN         ;increment tail (modulo BUFLEN)
39              cpx     TXHEAD          ;compare new tail to head
40              beq     PUTBERR         ;if equal ring is full
41              stx     TXTAIL          ;else update tail to save write
42              clc                     ;clear C to indicate success
43              bra     ENDPUTB         ;and goto finish
44  PUTBERR     sec                     ;set carry to indicate error
45  ENDPUTB     cli                     ;reenable interrupts
46              pulh
47              pulx
48              rts
49  ;------------------------------------------------------------------
50  ; SCI TX ISR
51  ;     if TXBUF not empty head of ring buffer is transmitted
52  ;------------------------------------------------------------------
53  SCITXISR    pshh                    ;h not auto stacked
54              clrh
55              ldx     TXHEAD          ;load buffer head
56              cpx     TXTAIL
57              beq     TBUFEMPTY       ;if head=tail then txbuf empty
58              lda     TXBUF,x         ;else get byte at head of ring
59              bsr     INCMODN         ;and advance head (modulo BUFLEN)
60              stx     TXHEAD
61              tst     SCIC2           ;load SCIC2 for clearing TDRE
62              sta     SCID            ;transmit byte, clear TDRE
63              bra     RTISCITX
64  TBUFEMPTY   bclr    TIE,SCIC2       ;disable TDRE interrupts
65  RTISCITX    pulh
66              rti
67  ;------------------------------------------------------------------
68  ; INCMODN(X)
69  ;     Increments X modulo BUFLEN  (i.e. X=(X+1) mod BUFLEN)
70  ;------------------------------------------------------------------
71  INCMODN     incx                    ;increment x
72              cpx     #BUFLEN
73              bne     ENDINCMODN      ;if not equal to BUFLEN do nothing
74              clrx                    ;else x=0
75  ENDINCMODN  rts
```

Code Listing 3.4: (*Continued*) Interrupt-Based Ring-Buffered SCI Simplex (Transmit) Driver.

(TDRE=0) the byte has to be queued and the code again branches to the QUEUEIT. Otherwise, the queue is bypassed and the data is stored to SCID and C is cleared for the return value before the code branches to the end of the subroutine.

Beginning at label QUEUEIT, on line 34, is where the subroutine attempts to queue the byte or return C=1 indicating that the queue is full. The first step is to enable SCI transmit interrupts (in case the queue is changing from the empty state). The data byte in A is then optimistically stored at the queue tail, although the write to the queue is not committed until the queue is determined to be not full (lines 38-40). If the queue is not full, TXTAIL is updated, the C flag is cleared, at which point the code branches to the end of the subroutine.

The SCI transmit interrupt service routine begins on line 53. The queue if first checked to see if it is empty, indicating that it had changed to the empty state the last time the ISR was triggered. If so, transmit interrupts are disabled on line 64 (to prevent the ISR from being continually triggered as described above) before the ISR returns. If the queue is not empty, the byte at the head of the queue is loaded on line 58 and the head is incremented (lines 59 and 60). If this makes the queue become empty, then the empty queue will be processed next time the ISR is triggered. The data loaded from the head of the queue is transmitted on line 62. In order for TDRE to be cleared, it must be read prior to this; this is done on line 61. The ISR subsequently returns.

3.2   SERIAL PERIPHERAL INTERFACE (SPI)

The serial peripheral interface is a bidirectional master-slave synchronous interface that is based on an 8-bit shift register. The SPI master generates a baud clock signal that is used by all SPI devices to coordinate data transfers. As shown in Figure 3.7, the peripheral device connected to the SPI also includes a shift register. Together, the two 8 bit shift registers are interconnected to form a 16 bit rotate left register. A data transfer consists of an 8 bit shift, which results in the exchange of data

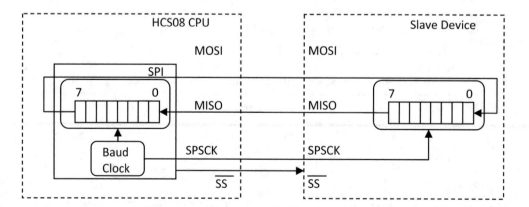

Figure 3.7: Connection from HSC08 SPI master to slave peripheral device.

between the master and slave devices. Many integrated circuits, such as digital to analog converters and EEPROM chips, have SPI interfaces. The SPI can also be used with a standard discreet logic shift registers to implement expanded I/O interfaces.

3.2.1 MC9S08GQ8 SPI

The MC9S08QG8 SPI interface consists of four pins: MOSI, MISO, SPSCK and $\overline{SS}$. The directions of these pins depend on whether the device is configured to act as master or slave and whether full-duplex or simplex communication is used. Interconnection of SPI devices consists simply of tying together the corresponding pins on devices. MOSI and MISO are the data input and output pins used to interconnect the master and slave shift registers. MISO (master in slave out) is an input pin for the master and output pin for the slave; MOSI (master out slave in) is an output pin for the master and input pin for the slave. SPSCK, the SPI baud rate clock, is output from the master to the slave. The clock polarity can be configured as rising or falling edge, with the clock edge occurring in the middle of the transmitted data bit. The clock is only active during an SPI transfer (for 8 SPSCK cycles while $\overline{SS}$ is active). The active-low slave select ($\overline{SS}$) signal is output by the master at the start of an SPI transfer to indicate to the slave that a transfer is active. This provides framing for the serial interface; the falling edge of $\overline{SS}$ marks the start of a frame, while the rising edge marks the end.

The SPI status, control and data registers are shown in Figure 3.8. SPE is the SPI enable bit; when clear, the SPI module is disabled to conserve power and the SPI pins can be used by other peripherals. Configuration bit LSBFE (lsb first enable) selects the bit direction of the transfer. Most-significant bit first (LSBFE=0) is the default mode of operation. The SPI baud rate is controlled by two clock dividers configured via the SPIBR register. The first is a clock prescaler, controlled the three bit SPPR field, which results in a clock whose frequency is $\frac{f_{bus}}{1+SPPR}$. The second is the primary clock divider, controlled by SPR, which further divides the prescaled clock by 2^{SPR+1}. Thus, the SCI baud rate is related to the bus clock and SPIBR bits by $f_{SPI} = \frac{f_{bus}}{(1+SPPR)(2^{SPR+1})}$ or $(1+SPPR)(2^{SPR+1}) = \frac{f_{bus}}{f_{SPI}}$. To determine the baud divisor, the bus clock to baud clock ratio must be factored into a number from 1 to 8 times a power of two up to 256. The minimum divisor is 2 and the maximum is 2048.

Example 3.2. Write the instruction(s) necessary to configure the SPI to operate at 200 kBd, assuming the bus clock frequency is 4 MHz.
Solution: $(1+SPPR)(2^{SPR+1}) = \frac{f_{bus}}{f_{SPI}} = \frac{f_{bus}}{200k} = 20 = 4 \times 5$. The only values that satisfy this exactly are SPPR=3, SPR=1. Thus, the value of the SPIBR register needs to be \$31.
Answer: mov #\$31,SPIBR

Since the baud clock is generated by the SPI master, it is not necessary to configure the baud rate of the slave devices. The SPI master clock frequency must fall within the specified timing range allowed by the slave. In addition, the master clock must be configured to match the timing

SPIC1: SPI Control Register 1 (memory mapped at address $0028)

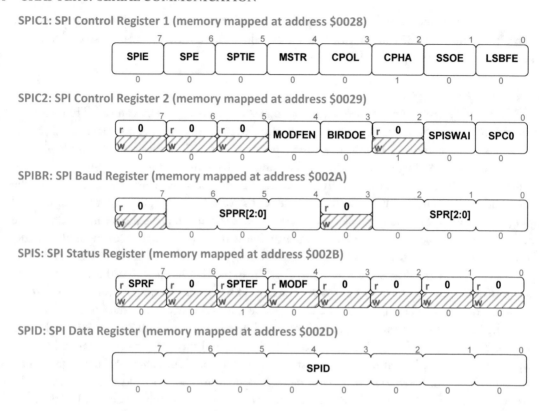

Figure 3.8: SPI control, status and data register formats.

requirements of the slave. There are two important considerations with respect to slave timing requirements: which edge of the clock the slave is using as the active edge and whether an inactive edge is required to precede each active edge. For SPI outputs, a data bit is shifted out on one edge of the clock (the shift edge) and sampled on the opposite edge of the clock (the sample edge). Two SPI configuration bits, CPOL and CPHA, control the phase and polarity of the SPI baud clock to configure these edges. Table 3.4 summarizes the clock characteristics controlled by these bits. In general, the four operating modes are incompatible and the master must be configured to match slave device timing (or vice versa). Each mode has been assigned a number, 0–3, defined by the two bit value CPOL:CPHA. The modes are also sometimes referred to by listing the (CPOL,CPHA) pair; for example, mode (0,1) refers to mode 1 or CPOL=0 and CPHA=1. CPHA controls which clock edge appears first after $\overline{SS}$ is asserted (note that $\overline{SS}$ is always asserted one-half baud clock period before the first edge). When CPHA=0, the sample edge is the first edge of the clock; otherwise, the shift edge appears first. Most devices that operate with CPHA=0. Note that when CPHA=1, $\overline{SS}$ can remain low during back-to-back transfers; CPHA=0 requires that $\overline{SS}$ go inactive between transfers. CPOL selects

Table 3.4: Summary of SPI clock polarity and phase configurations.

Mode Names	CPOL	CPHA	Sample Edge	Shift Edge	Initial Edge (after $\overline{SS}$ asserted)
0 or (0,0)	0	0	Rising	Falling	Sample
1 or (0,1)	0	1	Falling	Rising	Shift
2 or (1,0)	1	0	Falling	Rising	Sample
3 or (1,1)	1	1	Rising	Falling	Shift

the inactive edge of the clock to match the slave timing requirements. For some receive-only slave devices (master always outputs data to slave), the only consideration is the active edge of the clock; that is, the device can operate with either clock polarity as long as the sample edge polarity (rising or falling) matches the device requirements. For slave devices that transmit data to the master (or exchange data), a shift edge is usually required to precede a sample edge, giving the slave circuitry an opportunity to synchronize to the master clock.

The SPI duplexity and pin direction is controlled by the MSTR, SPC0, BIDIROE, MODFEN and SSOE pins. The operating mode of the SPI and the function of each pin in these modes are summarized in Table 3.5. Pins not used by SPI are available for general purpose I/O or one of the

Table 3.5: Duplexity and pin configuration options for the SPI.

MSTR	SPC0	BIDIROE	Operating Mode	SPSCK PIN	MOSI PIN	MISO PIN
1	0	-	Full Duplex Master	SPI Clock Output	SPI Data Output	SPI Data Input
1	1	0	Half-Duplex Master	SPI Clock Output	SPI Data Input	not used by SPI
1	1	1	Half-Duplex Master	SPI Clock Output	SPI Data Output	not used by SPI
0	0	-	Full-Duplex Slave	SPI Clock Input	SPI Data Input	SPI Data Output
0	1	0	Half-Duplex Slave	SPI Clock Input	not used by SPI	SPI Data Input
0	1	1	Half-Duplex Slave	SPI Clock Input	not used by SPI	SPI Data Output

other pin functions. MSTR (master mode select) selects whether the MC9S08QG8 SPI is operating as a SPI master (MSTR=1) or slave. The duplexity of the channel is subsequently selected by the SPC0 (SPI pin control) and BIRDIROE (bidirectional output enable) pins. SPC0=0 selects full-duplex modes. If the SPI is operating as master, the MISO is the data input and MOSI is the data output; in slave mode, the directions of these pins is reversed. A half-duplex (or simplex) channel is formed when SPC0=1. When configured this way, either the MOSI pin or MISO pin is used as the input/output pin depending, respectively, on whether the SPI is configured as master or slave. In either case, BIDIROE=0 configures the pin as input and BIRDIROE=1 configures it as output.

As shown in Table 3.6, the active low $\overline{SS}$ pin is always treated as an input when the SPI is operating as a slave. When master mode is configured, $\overline{SS}$ can either be disabled (pin not used by SPI), configured as a mode fault input, or used as the normal $\overline{SS}$ output. When a single slave is used, it may be possible to tie the $\overline{SS}$ pin low on the slave device; in this case the $\overline{SS}$ pin on the

Table 3.6: Slave select pin configuration options for the SPI.

MSTR	MODFEN	SSOE	SS Pin
0	-	-	Active Low SS Input
1	0	-	not used by SPI
1	1	0	Mode Fault Input
1	1	1	Active Low SS Output

MC9S08QG4/8 is not required for use by the SPI and can be disabled. When configured as a mode fault input, the $\overline{SS}$ pin can detect when another device is trying to become SPI master. When this happens, all SPI output pins are disabled and the MODF flag in SPIS becomes set.

There is a single interrupt vector for the SPI. The SPI status register (SPIS) has three flags, each of which has an associated interrupt enable in SPIC1. Flag SPRF indicates the SPI receiver is full. SPRF is cleared by reading SPIS followed by reading SPID. The associated interrupt enable, SPIE, can be used to send an interrupt request to the CPU when data is received in slave mode. MODF becomes set when a mode fault is detected (see above). MODF=1 interrupt requests are enabled by the SPIE (receive) interrupt enable. MODF is cleared by reading SPIS followed by writing SPIC1 (to reconfigure the device). SPTEF (SPI transmit buffer empty) is raised at the completion of an SPI transmit (or exchange) operation. The SPTIE interrupt enable bit allows SPTEF=1 to generate an SPI interrupt request. SPTIE is cleared by reading SPIS followed by writing SPID.

A write to SPID places the stored data in the SPI transmit register and initiates a SPI transfer if the SPI is configured as master. A read returns the value in the SPI receive register. The transmit and receive registers are double buffered. This means that while transmitting, a second value can be written to SPID before the previous transmission completes; furthermore, when receiving, a second byte can be in reception while a previously received byte is in the receive register. If this second byte reception completes before the previous byte is read from the receive buffer, an overrun condition results and this cannot be detected by software (there is no associated status flag); thus software should take this into account.

3.2.2   MC9S08QG8 SPI DRIVER EXAMPLES

The SPI is commonly used to interface a microcontroller to peripheral devices and is often selected over the SCI interface when higher baud rates are desired. In this section, two driver examples are provided to illustrate the use of the device for this purpose. Both drivers operate in master mode and use polled-I/O. The first driver is a generic 8-bit bidirectional SPI driver that provides both transmit and receive functions. The second is a 16-bit output only driver.

Code Listing 3.6 lists the memory mapped I/O register definitions related to the MC9S08QG8 serial peripheral interface. In addition to the five I/O registers described in Figure 3.8, the bit locations of the status flags SPRF and SPTRE are defined.

```
1   SPIC1        equ   $0028
2   SPIC2        equ   $0029
3   SPIBD        equ   $002A
4   SPIS         equ   $002B
5   SPID         equ   $002D
6   SPRF         equ   7
7   SPTRE        equ   5
```

Code Listing 3.5: SPI Driver Common Code.

3.2.3   FULL-DUPLEX 8-BIT POLLED-I/O SPI DRIVER

The driver in Code Listing 3.6 illustrates the basics of using polled-I/O to synchronize SPI transmit and receive operations when the SPI is operating as a master in full-duplex mode. The initialization subroutine INIT_SPI configures SPIC1 and SPIC2 to enable the SPI as a full-duplex master in mode (0,0). A baud rate divisor of $31 is then programmed, as computed in Example 3.1, to yield a Baud rate of 200 kBd. Finally, a dummy read of SPIS precedes a read of SPID to ensure that the initial state of SPRF is clear following initialization. Driver subroutine SPI_RXTX accepts as its only parameter a value in A to be transmitted, and returns back in A the byte received from the slave. After writing the transmit value to SPID to initiate the transfer, SPRF is polled until the transmit is complete. Because the transmission is guaranteed to finish within approximately 40 μs, there is no need to feed the watchdog. Finally, a load from SPID gets the received byte value in A and clears SPRF before returning.

3.2.4   SIMPLEX 16-BIT POLLED-I/O SPI DRIVER

Code Listing 3.7 illustrates the operations necessary to perform 16 bit transfers from the SPI to a slave. The driver operates the SPI in simplex mode with the channel direction set to output only. The primary difference in the driver initialization in Code Listing 3.7, compared to that of Code Listing 3.6, is that the dummy read sequence to clear SPRF is not necessary since SPRF is not used for polling by this driver. In addition, different configuration register settings yield half-duplex operation and illustrate the use of mode (1,1) and a higher baud rate of 2 MBd.

Driver subroutine SPI_TX16 accepts a single 16 bit parameter in HX, which contains the bytes to be transmitted (the MSB, H, is transmitted first). Accumulator A is callee-saved because it is modified by the subroutine. To transmit a byte on the SPI, the driver waits until the SPI transmit buffer is empty by polling the SPTEF flag. SPTEF=1 indicates that it is safe to write to SPID; there is no need to poll until the transmission is complete because no data is received from the slave. Even though the transmit register is double buffered, the SPTEF must be polled on both writes to SPID because the register may still contain data from a previous call to the driver subroutine.

```
1    ;*************************************************************
2    ;  SPI Initialization
3    ;*************************************************************
4    INIT_SPI    mov    #$52,SPIC1 ;enable SPI,full-duplex master
5                mov    #$10,SPIC2 ; w/ CPOL=CPHA=0,SS output enabled
6                mov    #$31,SPIBD ;Bd rate divisor:200kBd @ 4MHz bus
7                tst    SPIS       ;read SPIS for clearing SPRF
8                tst    SPID       ;read SPID to clear SPRF
9                rts
10   ;*************************************************************
11   ;  SPI_RXTX
12   ;     byte in A is transmitted on SPI;
13   ;     received byte returned in A
14   ;*************************************************************
15   SPI_RXTX    sta    SPID         ; write SPID  to start transfer
16   RXTXPOLL    brclr  SPRF,SPIS,RXTXPOLL ;poll until complete
17               lda    SPID         ;load RX data, clear SPRF
18               rts
```

Code Listing 3.6: Polled Full-Duplex SPI Master Driver.

```
1    ;*************************************************************
2    ;  SPI Initialization
3    ;*************************************************************
4    INIT_SPI    mov #$5E,SPIC1 ;enable as half-duplex, output MOSI
5                mov #$19,SPIC2 ; with CPOL=CPHA=1, SS output enabled
6                mov #$00,SPIBD ;baud divisor for 2MBd @ 4MHz bus
7                rts
8    ;*************************************************************
9    ;  SPI_TX
10   ;     16b value in HX transmitted on SPI (msb first);
11   ;*************************************************************
12   SPI_TX16    psha                        ;callee save
13   TXPOLL1     brclr  SPTEF,SPIS,TXPOLL1 ;wait until TX buffer empty
14               pshh                        ;transfer H to A
15               pula
16               sta    SPID                 ;write MSB to transmit buffer
17   TXPOLL2     brclr  SPTEF,SPIS,TXPOLL2 ;wait until TX buffer empty
18               stx    SPID                 ;write LSB to transmit buffer
19               pula                        ;callee restore
20               rts
```

Code Listing 3.7: Simplex (Output) SPI Master Driver with 16B Transfers.

3.3   INTER-INTEGRATED CIRCUIT (IIC)

The inter-integrated circuit (IIC or I^2C) bus is a two-wire synchronous bidirectional bus standard designed for low-throughput interconnection of multiple masters and peripherals. The organization of an IIC bus is illustrated in Figure 3.9. Because the bus can have multiple peripherals attached,

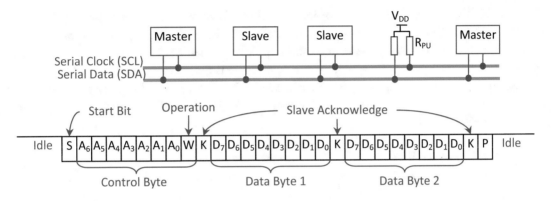

Figure 3.9: IIC bus organization and packet format.

each slave is assigned a unique 7bit address that is used to determine which is being accessed.

As a synchronous bus, IIC requires both clock and data signals. The serial clock line (SCL) carries the clock signal generated by a master accessing the bus. The serial data line (SDA) contains the data bits being transmitted; depending on the bus operation being performed, SDA can be driven by a master or a slave. To facilitate connecting multiple masters, SCL and SDA are open-drain lines requiring an external pull-up resistor to generate a logic high. In addition, a bus arbitration procedure is defined to resolve access conflicts when two masters try to access the bus simultaneously.

The frame format used in IIC is also shown in Figure 3.9. All data bits on SDA are sampled on the rising-edge of the clock signal on SCL, are required to remain stable when the clock is high, and may change when low. The exception is that start and stop bits are signaled by allowing SDA to change when SCL is high; a start bit is signaled by a falling transition of SDL when SCL is high, and a stop bit a rising transition. A basic frame consists of a control byte and one or more data bytes between a start and stop bit. The control byte consists of the 7 bit address of the slave being accessed along with one bit indicating whether a read from slave (1) or write to slave (0) is being performed by the master. The control byte and data bytes are transmitted msb first and each can be acknowledged by the receiver, which does so by signaling a low on SDA for one bit period following each byte. The acknowledge bit allows the transmitter to determine whether the byte was successfully transmitted.

The use of open-drain busses also facilitates arbitration and flow control. All masters monitor the bus and only attempt access when the line is idle (high condition on both lines between a stop and start bit). If a master is not driving SDA low but senses that it is low, then it knows another master is already using the bus and stops transmission; in this case the master that backed-off has lost arbitration. Similarly, if a master senses that SCL is low when it should be high, this indicates that a slave is stretching the clock to indicate that it is not yet ready to proceed with the next byte transfer, providing a basic flow control mechanism.

3.3.1   MC9S08QG8 IIC

The MC9S08QG8 IIC module can be operated in master or slave mode and uses two pins for SDL and SCL. The IICPS field in system options register SOPT2 determines the pins used for the SCI, which can be PTA2 and PTA3 for SDA and SCL, respectively, when IICPS=0 (the default) or PTB6 and PTB7, respectively, when IICPS=1. Since master mode is the more common role for the CPU, this section focuses on those aspects necessary to operate in master mode; details on slave mode are obtained from the device data sheet. In master mode, the IIC is accessed through four I/O registers, as shown in Figure 3.10.

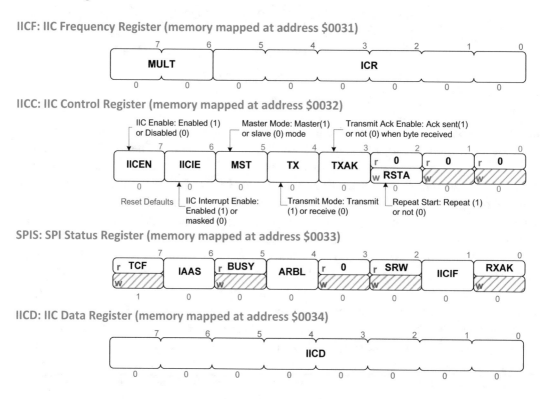

Figure 3.10: IIC control, status and data register formats.

IICF programs the baud rate and hold time of the module. The hold time is defined as the time after a falling clock edge when the data on SDA can change; some slave modules have specific hold time requirements that must be satisfied. The relationship between the MULT and ICR fields and the baud rate and hold time is complex and is listed in a table in the device data sheet. Since 100 kHz is a common baud rate and 4 or 8 MHz are common CPU bus clock frequencies, Table 3.7 lists the hold times for five IICF values that give a 100 kHz baud rate at 4 or 8 MHz bus clock frequencies.

Table 3.7: Hold times for different IICF values and clock frequencies at 100 kHz baud rate.

Bus Clock Frequency	IICF Value	Hold Time
4MHz	$0B	2.25 µs
4MHz	$40	0.32 µs
8MHz	$14	2.13 µs
8MHz	$4B	1.13 µs
8MHz	$80	0.88 µs

The IICC register is the primary control register for the IIC device. IICEN must be set to enable the IIC device. IICIE is the local interrupt mask, which must be set if IIC interrupts are being used. MST configures the device for active master mode and should only be set when the master is actively using the bus. When MST is changed from 0 to 1, a start bit is automatically signaled on the bus by the IIC interface; when changed back to 0, a stop bit is signaled (unless arbitration was lost). TX determines whether the IIC module is transmitting (1) or receiving (0) a byte. TXAK is set or cleared to specify the value that should be on SDA during the acknowledge bit period after the IIC module has received a byte. RSTA allows the master to do a repeated start so it does not have to relinquish the bus between transfers; this is also used in certain transfers to signal a change in operation (for example, a change from read to write).

IICS is the status register. The transfer complete flag (TCF) becomes set at the end of each byte transferred and is cleared automatically at the start of the next transfer or receive (via read or write to IICD). It can be used to poll for the end of a transfer or trigger an IIC interrupt. IAAS and SRW are used only for slave mode. BUSY is automatically set between start and stop bits on the bus; it can be used to determine if the bus is free (BUSY=0) or not before initiating a transmission. ARBL becomes set when this IIC master loses arbitration; it should be checked after each byte transferred to ensure that arbitration was not lost. IICIF is the IIC interrupt flag, which becomes set whenever TCF, ARBL, or IAAS change from 0 to 1. IICIF should be polled in master mode instead of TCF and ARBL because either of these events can end a transmission. RXAK holds the value of the acknowledge bit received from a slave device; it should be polled after each byte transmitted to ensure the byte was received by the slave.

A store to the IIC data register begins a transmission when MST=1 and TX=1 in IICC (master transmit mode). A read from IIC data register when MST=1 and TX=0 initiates a read from a slave. The IIC data register is not double-buffered and reads and writes must be issued at the right time for correct operation.

3.4   IIC DRIVER EXAMPLES

Like the SPI, IIC is a synchronous bus that can generally achieve higher baud rates than SCI. For the same baud rate, it achieves lower overall throughput compared to SPI due to the overhead of transmitting control bytes for each transfer and due the inclusion of start and stop bits. The flexibility of having multiple devices interconnected by just two wires is the primary advantage of IIC.

Many devices have protocols that require specific sequences of transfer operations (packet formats) to ensure correct operation. The drivers in this section utilize this principle to illustrate both generic use of the IIC as well as a specific case of interfacing an Atmel EEPROM.

Code Listing 3.8 provides memory-map definitions for the IIC module as well as common driver subroutines used by the driver examples. Since the IIC packet formats for different peripherals are composed of basic operations, driver subroutines are provided to handle these protocol operations. These include start bit signaling, master transmit, master receive and stop bit signaling.

All drivers in this section operate the IIC in master mode with a 100 kHz baud rate; thus, the initialization subroutine is the same and is included in the common code in Code Listing 3.8 rather than repeated in each subsequent code listing. Subroutine INIT_IIC begins by configuring the baud rate of the IIC; assuming an 8 MHz bus clock, the value $14 is written to obtain a 100 kHz baud rate, as indicated in Table 3.7. None of the drivers provided require the master to acknowledge received bytes; thus, the TXAK bit in IICC is set in the initialization subroutine to specify that a high level is to be driven on the SDA line during master acknowledge cycles. When not actively driving the IIC bus, the IIC module must be placed into slave mode (even if not acting as a slave). Thus, the MST bit in IICC is cleared next to ensure that the device is not in master mode. Having the TXAK bit set will also ensure that when in slave mode, if the slave address in the IICA register should ever match a slave address on the bus, the IIC module will not acknowledge. Finally, IICEN is set before return from the initialization subroutine to activate the IIC.

Driver subroutine IIC_START signals a start condition on the IIC bus. The subroutine returns C=0 if the master successfully acquires the bus and C=1 otherwise. The subroutine first uses a BRSET instruction to test the BUSY flag in IICS. In a multi-master setup, the bus can be busy if another master is using the bus. In a single master setup, the BUSY flag can also be set if a new transmission is started before the signaling of the stop bit from a prior transmission has completed. If BUSY=1, BRSET will set the carry flag and branch to the end of the subroutine, returning C=1. If BUSY=0, the branch falls through and the subroutine proceeds by clearing the ARBL flag and subsequently enabling master mode by writing a 1 to MST in IICC. This begins signaling a start bit on the bus.

Driver subroutine IIC_OUTB implements a master transmit of one byte on the IIC bus. It is used for both control and data byte transmission. The subroutine receives the value to be transmitted as a parameter in A and returns C=0 if the transmission was successful (acknowledged by the slave), and C=1 otherwise. The IIC module is first configured for transmit mode by setting TX in IICC. Recall that when TX is set, a write to IICD will initiate a transmission. This step is performed next with a store to IICD. A BRCLR instruction is subsequently used to poll until the IICIF flag in IICS becomes set, indicating that either the transmission completed or arbitration was lost. The BRCLR

```
1    IICA         equ    $0030     ;memory-map locations of I/O registers
2    IICF         equ    $0031
3    IICC         equ    $0032
4    IICS         equ    $0033
5    IICD         equ    $0034
6    RSTA         equ    2         ;flag and control bit locations
7    RXAK         equ    0
8    TXAK         equ    3
9    MST          equ    5
10   IICIF        equ    1
11   IICEN        equ    7
12   ARBL         equ    4
13   TX           equ    4
14   BUSY         equ    5
15   ;****************************************************************************
16   ;IIC Master Initialization
17   ;****************************************************************************
18   INIT_IIC     mov    #$14,IICF      ;set 100kHz baud rate (@8MHz clock)
19                bset   TXAK,IICC      ;master will not ACK
20                bclr   MST,IICC       ;not master mode
21                bset   IICEN,IICC     ;enable IIC
22                rts
23   ;****************************************************************************
24   ;IIC_START  Signal IIC Start; return C=1 if bus busy, else c=0
25   ;****************************************************************************
26   IIC_START    brset  BUSY,IICS,END_IICST ;check if bus free
27                bset   ARBL,IICC      ;clear ARBL flag
28                bset   MST,IICC       ;enable master mode to signal start
29   END_IICST    rts
30   ;****************************************************************************
31   ;IIC_OUTB   Implement Master transmit of one byte (after start)
32   ;           byte value passed in A, return C=0 If ACK received, else C=1
33   ;****************************************************************************
34   IIC_OUTB     bset   TX,IICC        ;configure mode as master transmit
35                sta    IICD           ;begin transmission of data byte
36                brclr  IICIF,IICS,*   ;wait for end of transmission or error
37                bset   IICIF,IICS     ;clear IICIF
38                brset  RXAK,IICS,ENDIICOB ;set C flag equal to ACK value
39   ENDIICOB     rts
40   ;****************************************************************************
41   ;IIC_INB    Implement Master receive of one byte (after start)
42   ;           returns byte in A, C=1 if IIC error
43   ;****************************************************************************
44   IIC_INB      bclr   TX,IICC        ;mode is RX
45                lda    IICD           ;dummy read to start receive data byte
```

Code Listing 3.8: IIC Master Driver Common Equates and Subroutines (*Continues*).

```
46                 brclr  IICIF,IICS,*    ;wait for end of transmission or error
47                 bset   IICIF,IICS      ;clear IICIF
48                 brset  ARBL,IICS,ENDIICIB; set carry if arbitration lost err
49  ENDIICIB       rts
50  ;*********************************************************************
51  ;IIC_STOP   Signal IIC Stop; return nothing
52  ;*********************************************************************
53  IIC_STOP    bclr   MST,IICC           ;change from MST=1 to =0 signals stop
54                 rts
```

Code Listing 3.8: (*Continued*) IIC Master Driver Common Equates and Subroutines.

uses the special branch target "*", which indicates that the branch target is the instruction itself. After falling out of the poling loop, IICIF is cleared, then the value of RXAK is loaded into the C flag using the BRSET instruction. Thus, the return value will be C=1 if the transmission was not acknowledged and C=0 otherwise.

Driver subroutine IIC_INB implements a master receive byte. It returns the received byte in A and C=0 if successful, or C=1 if arbitration was lost. After placing the module into receive mode, a load from IICD is used as a dummy read to initiate an IIC read cycle. The value returned by the load is not used; a later read to SCID after the byte has been received will return the received value. After the dummy load, IICIF is polled to determine when the receive operation completes. For a receive operation, arbitration lost is the only detectable error. Thus, before returning, BRSET is used to place the ARBL flag into the C flag as a return value.

Driver subroutine IIC_STOP signals a stop bit by returning the IIC module to slave mode.

3.4.1   GENERIC IIC DRIVER FOR DEVICES WITH SIMPLE READ/WRITE BEHAVIOR

Many IIC devices use simple IIC packet formats where a read from the slave returns one byte and a write sends one byte. The driver in Code Listing 3.9 can be used to provide the software side of the interface for these devices. The driver routines provide single byte transfer operations, called PUTBYTE (transmit to slave) and GETBYTE (receive from slave). The memory-map and common subroutines of Code Listing 3.8 are used by this driver.

Driver subroutine GETBYTE implements a full IIC packet transmission that includes a start signal, control byte, master receive and stop, all using calls to common driver subroutines. After each such call, an error is indicated by a return value of C=1. A branch is used after each call to test for this condition and branch to a common driver return point (END_IICGB) if true. The subroutine accepts the 7-bit slave address, left-aligned in index register X (bits 7 down to 1 contain the address). A call to IIC_START then signals a start condition on the bus. Upon return, if C=1 the bus was busy and a branch to END_IICGB will skip the remainder of the packet. If successful, the slave address in X is transferred to A and the lsb is set with an OR-mask to indicate a slave read operation. IIC_OUTB

```
;****************************************************************
;IIC_GETBYTE Implement Master receive of byte from slave whose address
;            is in X; return byte in A and C=0 If ACK received, else C=1
;****************************************************************
IIC_GETBYTE bsr IIC_START      ;implement IIC start
            bcs END_IICGB      ;if not busy, continue; else return C=1
            txa                ;address into A
            ora #$01           ;set lsb for slave read
            bsr IIC_OUTB       ;send slave address on bus
            bcs END_IICGB      ;if no ACK, return C=1
            bsr IIC_INB        ;else get byte from slave;
END_IICGB   bsr IIC_STOP       ;signal stop bit (or exit if error)
            bset TX,IICC       ;to TX so IICD read does not start receive
            lda IICD           ;get received data for return
            rts
    ;****************************************************************
;IIC_PUTBYTE Implement Master transmit of byte in A to slave whose address
;            is in X; return C=0 If ACK received, else C=1
;****************************************************************
IIC_PUTBYTE psha               ;callee save
            pshx
            bsr IIC_START      ;implement IIC start
            bcs END_IICGB      ;if not busy, continue; else return C=1
            txa                ;address into A
            lsra               ;clear lsb to signal master write
            lsla
            bsr IIC_OUTB       ;send slave address on bus
            bcs END_IICPB      ;if no ACK, return C=1
            lda 2,SP           ;get callee-saved data byte into A
            bsr IIC_OUTB       ;send byte to slave; C=1 if not ACK else 0
END_IICPB   bsr IIC_STOP       ;stop bit (or exit master mode if error)
            pulx               ;callee restore
            pula
            rts
```

Code Listing 3.9: Generic IIC PUTBYTE and GETBYTE Driver.

is then called to place the IIC control byte on the bus. If successful, a subsequent call to IIC_INB initiates the reception of one byte from the addressed slave.

END_IICGB is the common subroutine return point. It is executed whether or not there is an error. None of the operations performed starting at END_IICGB modify the C flag. Thus, the value of C that led the subroutine to this point also serves as the return value. In addition, the call to IIC_STOP, which signals a stop condition, is safe to execute for every case that leads the subroutine to END_IICGB. If arbitration is lost, master mode is automatically disabled (MST=0) by the IIC unit; since IIC_STOP clears MST, no stop signal is actually signaled in this case. For all other situations, a stop is signaled to end the packet transmission. So that the load of the received byte will not initiate another IIC transfer, the IIC mode is switched to TX before loading the received byte value into A. However, the value loaded is only valid if C=0.

IIC_PUTBYTE uses a similar sequence of operations to perform a slave write. The subroutine accepts the slave address in X (left-aligned) and the byte to be transmitted in A. After callee-saving A and X, IIC_START is called to signal the start of a packet. Then, the address in X is transferred into A as the parameter to send to IIC_OUTB for the control byte transmission. The address is logically shifted right then left to clear the lsb, which must be 0 in the control byte to signal the slave write. The remainder of the subroutine is similar to IIC_GETBYTE, with the primary difference that IIC_INB is called to receive a byte of data from the addressed slave.

3.4.2   DRIVER FOR AN ATMEL AT24C02B EEPROM INTERFACED VIA THE IIC BUS

The Atmel AT24C02B EEPROM is a 256 Byte EEPROM that is interfaced via IIC. The device has several read and write modes that allow access to one or more bytes per IIC packet. The packet formats used by the EEPROM single-byte read and write operations (also called random read and write) are shown in Figure 3.11. For an EEPROM write, the address of the byte stored is transmitted

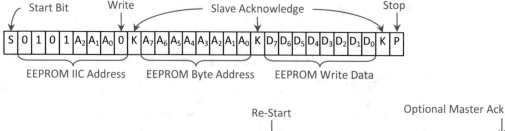

Figure 3.11: AT24C02B EEPROM single byte random write and read IIC sequences.

following the standard IIC slave address control byte. Three pins on the EEPROM device allow the three least significant slave address bits to be customized; the four most significant bits are fixed at 1010. After sending the byte address, the value to be stored is placed on the bus by the master. An EEPROM random read is more complex, requiring an interrupted write sequence, followed by a restart, followed by a read. As shown in Figure 3.11, the random read packet format follows the write packet format through the transmission of the address of the byte to be written; a restart is then signaled on the bus, and the EEPROM is addressed again using a standard slave read sequence.

Code Listing 3.10 lists the driver code required to perform AT24C02B random read and write operations. Subroutine EEPROM_READ takes the address of the EEPROM as a parameter, passed

```
;*********************************************************************
;EEPROM_READ: Random read of byte at address A, from EEPROM slave whose
;           address is in X; return byte in A and C=0 on success, else C=1
;*********************************************************************
EEPROM_READ pshx                ;callee save
            bsr  IIC_START       ;implement start bit
            psha                 ;save byte address
            lda  2,SP            ;get (callee-saved) slave address
            and  #$FE            ;clear lsb (slave write)
            jsr  IIC_OUTB        ;implement address control cycle
            bcs  END_EEPR        ;error in address cycle
            pula                 ;get saved byte address
            jsr  IIC_OUTB        ;output it
            bcs  END_EEPR        ;check for ACK
            lda  1,SP            ;get (callee-saved) slave address
            ora  #$01            ;set lsb for slave read
            bset RSTA,IICC       ;signal a rs-start
            bsr  IIC_OUTB        ;implement address control cycle
            bcs  END_EEPR        ;error in address cycle
            bsr  IIC_INB         ;receive value returned by EEPROM
END_EEPR    bsr  IIC_STOP        ;stop bit (or exit master mode if error)
            bset TX,IICC         ;to TX so IICD read does not start receive
            lda  IICD            ;get received data for return
            pulx
            rts
;*********************************************************************
;EEPROM_WRIT: Random write of byte value in A, to EEPROM address H, to
;     EEPROM slave whose address is in X; return C=0 on success else C=1
;*********************************************************************
EEPROM_WRIT pshx                ;callee save
            pshh
            psha
            bsr IIC_START        ;signal IIC start
            lda 3,SP             ;get (callee-saved) slave address
            and #$FE             ;make sure slave address set to write
            bsr IIC_OUTB         ;send slave address to IIC bus
            bcs END_EEPW         ;return C1 if not ACK
            lda 2,SP             ;get EEPROM address to be written
            bsr IIC_OUTB         ;send it on bus
            bcs END_EEPW         ;return C1 if not ACK
            lda 1,SP             ;get byte to be written
            jsr IIC_OUTB         ;send it on bus
END_EEPW    bsr IIC_STOP         ;stop bit (or exit master mode if error)
            pula                 ;callee restore
            pulh
            pulx
            rts
```

Code Listing 3.10: IIC EEPROM Driver.

in X, and the address of the byte to be read, in A, and returns the value read in A. The carry flag is set upon return if an error occurred. The subroutine EEPROM_WRIT takes as parameters the EEPROM slave address in X, the EEPROM memory address to be written in H, and the value to be written in A; it returns C=0 on success and C=1 otherwise. These subroutines follow the same basic structure as the IIC_GETBYTE and IIC_PUTBYTE from Code Listing 3.9, except for argument handling and the extra IIC cycle needed to send the EEPROM byte address.

EEPROM_READ first calls IIC_START to signal a start condition. It then sends the EEPROM slave address with the lsb clear, indicating that it will be starting a write cycle. The slave address cycle is followed by another call to IIC_OUTB to send the EEPROM byte address, then a re-start is signaled, as per the packet format specified in Figure 3.11. After the re-start, the EEPROM slave address is resent, this time with the lsb set to indicate a slave read. IIC_INB is then called to get the byte value returned from the EEPROM. Error handling is similar to that used in IIC_GETBYTE and IIC_PUTBYTE (Code Listing 3.9), and parameter handling is documented in the comments.

EEPROM_WRITE follows a similar sequence, except that no restart is required. After sending the EEPROM byte address, IIC_OUTB is instead called with the value to be stored into the EEPROM. It should be noted that an EEPROM write operation takes up to 5 ms to complete after the value is transmitted over the IIC bus. While delay synchronization could be used to ensure that the write to the EEPROM is complete before writing another value, the return value of the EEPROM_WRIT subroutine can be used instead. If an EEPROM write is initiated less than 5 ms before the previous write completes, the packet transmission to the EEPROM will fail and EEPROM_WRIT will return C=1. Software can retry the write operation by repeatedly calling EEPROM_WRIT until C=0. Thus, whenever calling EEPROM-WRIT software must check the carry flag to determine if the operation completed.

3.5   CHAPTER PROBLEMS

1. How does a synchronous serial communications interface differ from an asynchronous interface?

2. Sketch an oscilloscope view of an SCI transmission of the byte $3A, with one start bit, one stop bit, and even parity.

3. Figure 3.12 shows a view of one frame of a standard serial transmission, as it would appear on an oscilloscope, using 8 data bits, one start bit and even parity. Is there a parity error?

Figure 3.12: View of one frame of a standard serial transmission.

4. Which serial interface type has the has the highest data throughput? The lowest?

5. Which serial interface type(s) support multiple peripherals?

6. Sketch an oscilloscope view of a IIC read of the byte from an ATMEL AT24C02B EEPROM with slave address $A8. Assume the byte address is $7F and the value returned is $44.

7. Show how the SPI interface can be used with an external shift register to form an additional 8-bit output port.

8. Show how the SPI interface can be used with an external shift register to form an additional 8-bit input port.

9. Determine the baud rate divisor for the SCI necessary to achieve 19200 Bd with a 4 MHz system clock.

CHAPTER 4

Real-Time I/O Processing

Many embedded systems require that I/O operations be repeated periodically. Sometimes, the repetition applies to the main processing loop of the embedded system, while other times only part of the processing involves repetitive operations. Real-time I/O processing is about performing these repetitive tasks. The MC9S08QG4/8 processors have three modules that facilitate real-time processing. These are the real-time interrupt, which is well suited to forming a repetitive main loop or performing periodic tasks; the modulo-timer, which is useful for creating delay loops; and the pulse-width modulator, which is useful for creating delay loops or generating pulse-width modulated signals.

4.1   REAL-TIME INTERRUPT

The real-time interrupt (RTI) module generates periodic interrupt requests. It is based on a free-running counter that generates an interrupt request whenever it times out. Periodic interrupts are useful to perform periodic I/O operations, to support task switching, and in the implementation of real-time embedded operating systems. The real-time interrupt module on the HCS08 can be configured to operate from an internal 1 kHz reference source or from the external reference clock of the internal clock source (ICS).

4.2   MC9S08QG4/8 REAL-TIME INTERRUPT MODULE

The RTI module has a single status and control register, shown in Figure 4.1. The RTCLKS selects

SRTSC: System Real-Time Status and Control Register
(memory mapped at address $1808)

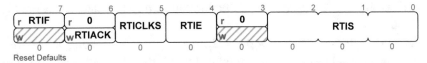

Figure 4.1: Format of the system real-time interrupt status and control register.

the clock source of the RTI module; when clear, an internal 1 kHz oscillator is selected; when set, the ICS external reference clock is used. According to the datasheet, the 1 kHz internal source can be off by as much as ±30%. RTIS controls the period (frequency) of the RTI. When RTIS=000, the

RTI is disabled. For any other value, the period between interrupts is $4\text{ ms} \times 2^{\text{RTIS}}$; thus, using the 1 kHz clock source, the RTI period can be set to seven progressively larger values in the range from 8 ms (RTIS=001) to 1.024 s (RTIS=111).

RTIF becomes set each time the RTI timer expires. RTIE enables the RTI interrupt request when RTIF=1. Thus, either polling or interrupt synchronization can be used. RTIACK is used to clear RTIF (by writing a 1 to RTIACK) and thus acknowledge the RTI interrupt.

4.2.1   PERIODIC SYSTEM WAKEUP USING THE REAL TIME INTERRUPT

Embedded systems are often tasked with periodically performing a necessary operation while remaining "idle" during the interval between tasks. The real-time interrupt module can be configured to maintain the timing of the desired inter-task period, allowing the embedded system designer to focus on programming the task to be completed. In addition, the microcontroller can be programmed to enter stop mode between tasks, minimizing the power consumed.

Code Listing 4.1 provides an example of a driver that uses the RTI module to generate a programmable real-time period that is up to approximately 65535 seconds (roughly 18 hours). The driver uses two global variables, RTICOUNT and RTIFLAG, to communicate with software. RTICOUNT maintains the number of RTI periods that have elapsed since the last time RTIFLAG was last set. When RTICOUNT reaches the desired number of periods (defined by constant RTIPERIOD) it is reset and RTIFLAG is set. Software can use RTIFLAG to poll for the end of the desired period and perform the necessary actions, and must clear it to detect the next period.

INITRTI initializes the RTI module to use the internal 1 kHz clock source with a RTI interrupt period of 1.024 s and enables RTI interrupts. The subroutine initializes all shared global variables before returning.

RITISR is the interrupt service routine, which will execute approximately every 1.024 s. The subroutine starts by acknowledging the interrupt request by writing a 1 to the RTIACK bit in RTISCI. Because RTISCI is not in the direct page, a BSET instruction cannot be used; instead, an OR-mask operation is performed. Then, the driver global variable RTICOUNT is loaded into HX, incremented, and compared to RTIPERIOD-1. If RTICOUNT is less than RTIPERIOD-1, then RTICOUNT is updated and the ISR returns; otherwise, RTICOUNT is cleared and RTIFLAG is set to indicate that the desired amount of time has elapsed. A longer interrupt period can be implemented by maintaining two or more "nested" counters.

Code Listing 4.2 demonstrates how the RTI driver from Code Listing 4.1 can be used in a modified main system loop. The main loop is modified to test the RTIFLAG at the top of the loop to see if the desired time has elapsed. If it has, the branch to ENDMAINLP is not taken and the main body of the loop is executed after RTIFLAG is cleared. Because RTIFLAG is the only global variable used and it is accessed and modified with a single CLR instruction, there is no need to make the access atomic by disabling interrupts. At the end of the main loop, the watchdog is fed and the CPU is placed in stop mode. Recall that stop mode must be enabled by setting the STOPE bit in the system options register SOPT1. Every 1.024 s, the RTI interrupt wakes the CPU from stop mode, and the

```
1   SRTISC      equ    $1808 ;memory-map location of SRTISC
2   RTIPERIOD   equ    2     ;number 1.024s periods between wake-ups
3   ;--------------------------------------------------------------
4   ;Required Global Variables
5   ;RTICOUNT    ds.w   1     ;number of 1.024 s periods elapsed
6   ;RTIFLAG     ds.b   1     ;set at end of each RTIPERIOD periods
7   ;--------------------------------------------------------------
8   ;RTI Module Initialization
9   INITRTI     psha              ;callee save
10              clra              ;initialize global variables
11              sta   RTIFLAG     ;RTIFLAG is clear
12              sta   RTICOUNT    ;RTICOUNT=0
13              sta   RTICOUNT+1
14              lda   #$57        ;RTIS=111 (period=1.024s), RTIE=1,
15              sta   SRTISC      ;and RTICLKS=0 (select internal clock)
16              pula              ;callee restore
17              rts
18  ;--------------------------------------------------------------
19  ;RTI Module Interrupt Service Routine
20  RTIISR      pshh              ;H not automatically stacked
21              lda   SRTISC      ;get current RTI status and control
22              ora   #%01000000 ;set RTIACK
23              sta   SRTISC      ;write to RTIISR to acknowledge
24              ldhx  RTICOUNT    ;get current period number
25              aix   #1          ;increment it
26              cphx  #(RTIPERIOD-1) ;compare to desired wake-up
27              blo   NOTTIME     ;if not wakeup time, goto end of ISR
28              lda   #1          ;else set Boolean global flag
29              sta   RTIFLAG
30              ldhx  #0          ;and write zero to current period
31  NOTTIME     sthx  RTICOUNT
32              pulh              ;restore H before return
33              rti
```

Code Listing 4.1: Real-Time Interrupt Driver to Perform Periodic System Wake-Up.

main loop is executed. For most of these wakeup periods, the CPU will simply test RTIFLAG and stop again; only when RTIFLAG is 1 will the main loop body execute. This method allows the CPU to minimize power consumption between loops.

4.3   MODULO TIMER MODULE (MTIM)

The modulo timer module (MTIM) is based on an 8 bit free-running modulo-M up-counter with programmable clock source and modulo value. The counter increments from 0 to one less than the modulo value (M), setting a flag upon overflow (the transition from M-1 to 0) and optionally triggering an interrupt; the counter then continues counting from 0. The MTIM counter can be used

```
1.   MAINLOOP:   tst    RTIFLAG    ;test if desired time has elapsed
2.               beq    ENDMAINLP  ;goto end of main loop if not
3.               clr    RTIFLAG    ;clear for detection of next period
4.   MAINBODY:   nop               ;replace with main loop body
5.   ENDMAINLP   sta    WATCHDOG   ;feed the watchdog
6.               stop              ;place CPU in stop mode until next IRQ
7.               bra    MAINLOOP   ;upon wake-up from stop, repeat
```

Code Listing 4.2: Using the Real-Time Interrupt Driver to Implement a Periodic Main Loop.

as an alternative to software delay loops to perform delay synchronization or as an alternative source of a periodic interrupt. MTIM operates in wait mode but not stop mode.

MTIMSC: Modulo Timer Status and Control Register (memory mapped at address $003C)

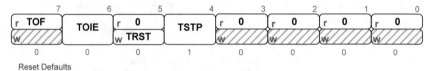

MTIMCLK: MTIM Clock Configuration Register (memory mapped at address $003D)

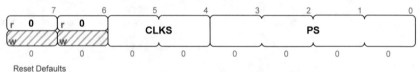

MTIMCNT: MTIM Count Register (memory mapped at address $003E)

MTIMMOD: MTIM Modulo Register (memory mapped at address $003F)

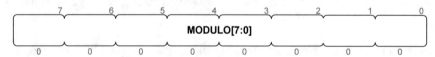

Figure 4.2: Modulo timer register formats.

The modulo timer is controller by 4 registers, as shown in Figure 4.2. MTIMSC is the status and control register for the MTIM. The TOF bit is the timer overflow status flag, which becomes set each time the counter rolls over to 0. TOF is cleared in one of three ways: by writing a zero to it after a read from MTIMSC, by writing to MTIMMOD, or when TRST is set. TOIE is the interrupt enable bit; a timer overflow interrupt request is issued when TOF=1 only if TOIE is set. TRST is the timer reset control bit; writing a 1 to TRST causes the counter to reset to $00 and TOF to clear. TSTP is the counter stop bit; when TSTP=1, the counter stops counting; counting resumes from the stopped value when TSTP is subsequently cleared (unless the counter is reset).

The MTIMCNT register contains the current counter value; software can read MTIMCNT to determine how much time has elapsed. MTIMMOD is programmed with the desired modulo value. A number from 1 to 255 sets the modulo value to that number, generating a period of MTIMMOD+1; setting MTIMMOD to 0 defines a modulo-256 (free-running) counter.

The MTIMCLK register controls the counter frequency. CLKS selects the counter clock source (00: bus clock; 10: TCLK pin falling edge; or 11: TCLK pin rising edge). PS defines the clock prescaler, which can be a number from 0 to 8; the MTIM counter frequency is defined as the MTIM clock frequency divided by 2^{PS}. For example, with a 4 MHz bus clock and PS=8, the MTIM clock frequency would be 15.625 kHz (MTIM period of 64 μs).

4.3.1 GENERATING DELAYS WITH THE MTIM

Among other uses, the modulo timer can be used to generate fixed software delays by programming the appropriate period in the modulo register, restarting the counter, and polling until the timer overflows. Code Listing 4.3 shows an example of a subroutine that delays for 1 ms at a 4 MHz bus clock. This approach is slightly easier than using software delay loops, as shown in Code Listing 1.3.

```
1    MTIMMOD      equ   $003F          ;MTIM memory map definitions
2    MTIMCLK      equ   $003D
3    MTIMSC       equ   $003C
4    ;-----------------------------------------------------------------
5    ;DELAY1US:  Delay for N us.  N is passed in A.
6    ;-----------------------------------------------------------------
7    DELAY1MS     mov   #$04,MTIMCLK   ;MTIM clock=(Bus Clk)/16
8                 mov   #250,MTIMMOD   ;write desired number of periods
9                 bset  5,MTIMSC       ;reset counter and TOF
10                bclr  4,MTIMSC       ;start counter
11                brclr 7,MTIMSC,*     ;wait for TOF
12                bset  4,MTIMSC       ;stop the counter
13                rts                  ;callee restore
```

Code Listing 4.3: 1 MS Delay Subroutine using the MTIM.

The subroutine configures the MTIM to use the bus clock divided by 16, which results in a 256 kHz clock, or a 4μs per period. 250 periods of this clock is 1 ms, so the MTIM modulo register is programmed to with the value 249 (recall the modulo value is one less than number of periods counted). After resetting the counter and starting the MTIM, the timer overflow flag TOF is polled to detect the end of the 1 ms period. After stopping the counter, the subroutine returns. The delay generated is slightly greater than 1 ms as the subroutine call overhead and MTIM configuration instructions take a small amount of time.

4.3.2   NON-BLOCKING SOFTWARE DELAYS USING THE MTIM

Often, tasks need to be performed periodically. While it was shown in Code Listing 4.1 that the RTI module can be useful to perform periodic processing, it has several limitations. The RTI can only be programmed to generate one of seven different delay values from 8 ms to 1.024 s, while the MTIM can be programmed to generate 2040 delay values (8 prescale values * 255 modulo values), from as short as one bus clock cycle to as long as 65536 bus clock cycles (about 16 ms at 4 MHz bus clock). In addition, while the RTI is useful for controlling periodic execution of the main system loop, as was shown in Code Listing 4.2, it is sometimes necessary to perform periodic processing of a smaller task for a limited period of time (for example, periodic sampling of a finite number of ADC values).

Delay synchronization can be used to provide periodic processing when the processing loop body executes in a constant number of clock cycles. However, processing often contains control flow changes that can make the processing time variable. In addition, interrupts can add variable delays even to precisely timed I/O sequences. In these cases, getting precise timing can require varying the amount of delay inserted between tasks, as shown in Figure 4.3 (top), which shows a timeline of three executions of a periodic I/O processing loop. Because the processing time of each iteration varies, the amount of delay required must be adjusted. Because the CPU blocks during each wait, it cannot overlap the delay generation and I/O processing.

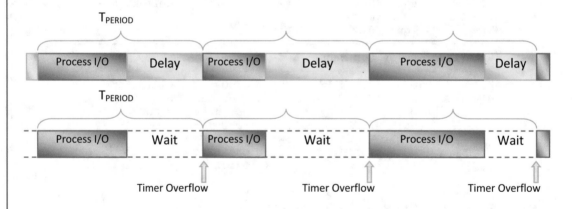

Figure 4.3: Periodic processing with blocking delays (top) and non-blocking delays (bottom).

In contrast to inserting variable delays between iterations, the MTIM counter runs in parallel with CPU processing and, after processing, software can wait until the end of the period to ensure precise timing. With this approach, shown in Figure 4.3 (bottom), each I/O processing period can absorb as much time as it needs during the given period. When the I/O processing is complete, software can poll (wait) until the end of the period, consuming the balance of the difference between the period and the processing time. The delays are non-blocking because the CPU can process the I/O in parallel with the timer. This approach offers simplicity over using software delays, even when the I/O processing time is not variable, because the I/O processing time does not need to be known. In addition, interrupts that occur during the processing interval will not affect the processing period unless they occur just before the timer overflows, delaying the time at which TOF is detected. This effect will introduce less jitter between iterations than using software delays, and will not affect the processing rate.

Code Listing 4.4 provides an MTIM driver that can be used to provide non-blocking delays for periodic I/O processing. Driver initialization subroutine MTIMINIT sets up the MTIM clock and

```
1    MTIMMOD      equ   $003F          ;MTIM memory map definitions
2    MTIMCLK      equ   $003D
1    MTIMSC       equ   $003C
2    ;--------------------------------------------------------------
3    ;MTIM Module Initialization
4    MTIMINIT     mov   #$04,MTIMCLK ;MTIM clock=(Bus Clk)/16
5                 mov   #250,MTIMMOD ;write desired number of periods
6                 rts
7    ;--------------------------------------------------------------
8    ;Resets and Starts the MTIM Counter
9    MTIMSTRT     bset  5,MTIMSC       ;reset the MTIM counter
10                bclr  4,MTIMSC       ;start the counter
11                rts
12   ;--------------------------------------------------------------
13   ;Stops the MTIM Counter
14   MTIMSTOP     bset  4,MTIMSC       ;stop the counter
15                rts
16   ;--------------------------------------------------------------
17   ;Waits until end of period
18   WAITTICK     brclr 7,MTIMSC,WAITTICK   ;wait for end of period
19                bclr  7,MTIMSC   ;clear overflow
20                rts
```

Code Listing 4.4: Non-Blocking Delays for Periodic Processing with the MTIM Module.

modulo value for a period of 1 ms (1 kHz processing loop rate). Subroutine MTIMSTRT is used to reset and start the MTIM counter when I/O processing is to begin. MTIMSTOP can be used to stop the MTIM counter when the periodic I/O processing is complete, to save power.

Driver subroutine WAITTICK waits until the MTIM counter overflows by polling for TOF=1. At that point, TOF is cleared for the next period and the subroutine returns. Even though WAITTICK is a blocking subroutine, the delays generated are considered non-blocking because the timer is running before WAITTICK is called (the remainder of the period is available for processing).

Code Listing 4.5 shows an example of using non-blocking delays to perform periodic sampling of N values from the ADC. The code sequence writes the N ADC samples to an array called ARRAY. MTIMINIT is called to set up the MTIM period. After initializing the loop counter, MTIMSTART is called to start the MTIM counter, marking the beginning of the first period. In the I/O processing loop, GETADC is called to sample one value from the ADC; one of the ADC drivers from Chapter 2 is required to provide subroutine GETADC. Subsequently, the ADC output is stored in the array, the loop counter update is done, and a call to WAITTICK is made to wait until the end of the sample period. Upon return from WAITTIC, the loop branch either starts another sample or terminates the loop.

```
1                  bsr   MTIMINIT   ;initialize the timer
2                  ldhx  #0         ;initialize loop counter
3                  bsr   MTIMSTRT   ;start MTIM
4      LOOP:       bsr   GETADC     ;get an ADC sample
5                  sta   ARRAY,X    ;store it in ARRAY[HX]
6                  aix   #1         ;increment loop counter
7                  bsr   WAITTICK   ;wait for end of period
8                  cphx  #NUMSAMPLES ;check if reached desired number
9                  bne   LOOP       ;repeat until
10                 bsr   MTIMSTOP   ;stop the timer
```

Code Listing 4.5: Using Non-Blocking Delays for Periodic ADC Sampling with the MTIM.

4.4   PULSE WIDTH MODULATION

The timer/pulse-width modulator module (TPM), like the MTIM, is based on a free-running counter with configurable clock and period. However, in addition to the functionality provided by the MTIM, the TPM supports multiple channels, each with an associated modulo register and I/O pin, that can be used to output pulses with programmable position, polarity and duration; to output periodic waveforms with programmable polarity, frequency and duty cycle; or to capture the time of occurrence of input events (rising or falling edges). This section focuses on the use of the TPM module for basic pulse-width modulation, in which each channel outputs a periodic digital waveform with a programmable polarity, frequency and duty-cycle.

4.5   MC9S08QG4/8 TPM

The MC9S08QG4/8 TPM is based on a free-running 16-bit up/down counter. The TPM I/O registers that are used for pulse-width modulation (PWM) are summarized in Figure 4.4. Only those features used for basic pulse-width modulation are covered.

The TPM status and control register is similar to that of the MTIM. TOF, TOIE, PS, and CLKS have the same purpose as the MTIM fields of the same name. When CLKS=01, the TPM clock is the bus clock divided by 2^{PS}; CLKS=00 stops the TPM clock, disabling the module. TOF is the overflow flag, which generates a TPM interrupt request when it becomes set as long as TOIE=1. CPWMS is not used for basic pulse-width modulation.

Each TPM channel has its own configuration and status register: TPMCnSC, where n is 0 or 1. For PWM operation, the CHnF flag indicates that the end of the active pulse period has been reached. ChnIE is the TPM CHn interrupt request enable (each channel has its own interrupt vector).

TPMSC: TPM Status and Control Register (memory mapped at address $0040)

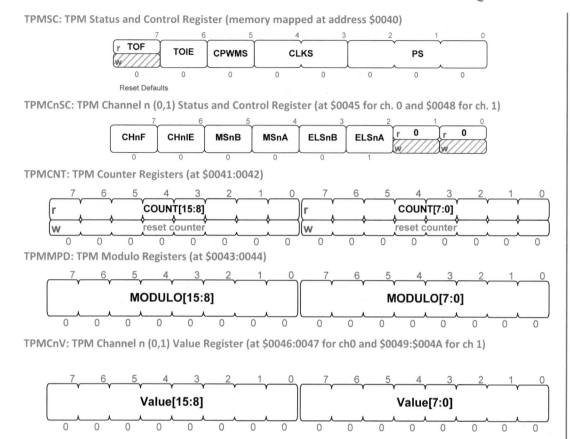

Figure 4.4: TPM register formats.

When set, an interrupt request is generated whenever CHnF=1. For PWM, MSnB:MSnA=10. When ELSnB:ELSnA=10, the pin TPMn has positive-polarity. Likewise, when ELSnB:ELSnA=01, pin TPMn has negative polarity.

TPMCNT is the 16-bit TPM counter register. A read returns the current value of the TPM counter and a write to either byte resets the counter. TPMMOD is the 16-bit modulo register, which defines the period of both PWM channels, indicating that the two PWM output signals must have the same period. TPMCnV is the channel-n value register. This value defines the duty-cycle of the active period of the PWM signal, relative to the TPMMOD register. When TPMCnV $\leq$ TPMMOD, the duty cycle is 100*TPMCnV/(TPMMOD+1); otherwise, the duty cycle is 100%.

4.5.1   TPM VARIABLE DUTY CYCLE DRIVER

Code Listing 4.6 provides an example of generating a variable duty cycle PWM waveform on pin TPM0 (TPM channel 0 pin). The TPM clock is selected to be the bus clock divided by 4 (PS=2), which gives $1\mu s$ per TPM clock period given a 4 MHz bus clock. The period of the PWM signal is statically configurable to be any integer from 2 to 65534, which provides frequencies from 500 kHz to

```
1   TPMSC        equ   $0040    ;TPM memory-mapped I/O register locations
2   TPMCNTH      equ   $0041
3   TPMMODH      equ   $0043
4   TPMC0SC      equ   $0045
5   TPMC0VH      equ   $0046
6   PWMPERIOD    equ   1000     ;PWM Period in us(between 2 and 65534)
7   ;-----------------------------------------------------------------
8   ;Initializes the the TPM Counter to generate a PWM signal on TPM0
9   ;with a period of PWMPERIOD and initial duty cycle of 0%
10  INITTPM      mov   #$02,TPMSC       ;PS=2 (divide 4); CLKS=00 stops tpm
11               ldhx  #(PWMPERIOD-1)   ;write period to TPMMPD
12               sthx  TPMMODH          ;write new modulo value
13               clc
14               clrx                   ;initial duty cycle to 0%
15               bsr   SETDUTY
16               mov   #$24,TPMC0SC     ; active high, edge-aligned PWM
17               bset  3,TPMSC          ;CLKS=01 (bus clock), start counting
18               rts
19  ;-----------------------------------------------------------------
20  ;Dynamically changes the duty cycle; duty cycle parameter is fixed
21  ;point fraction C.X, where C is carry flag and X is fraction part.
22  ; passed in X.  Returns nothing.
23  SETDUTY      ais   #-3              ;local storage space (DUTY,FRACDUTY)
24               stx   3,SP             ;initialize FRACDUTY
25               bcc   NOT_100          ;if carry set, then request is 100%
26               ldhx  #$FFFF           ;special case when 100% duty cycle
27               bra   ENDSETDUTY
28  NOT_100      ldhx  TPMMODH          ;get current modulo value (period-1)
29               aix   #1               ;PERIOD=modulo+1
30               sthx  1,SP             ;initialize DUTY=PERIOD
31               lda   2,SP             ;get DUTY LSB
32               ldx   3,SP             ;get FRACDUTY
33               mul                    ;multiply
34               stx   2,SP             ;store product integer part to DUTY LSB
35               rola                   ;msb of product fraction into C for round
36               clra                   ;clear A
37               adc   2,SP             ;add rounding bit to DUTY LSB
38               sta   2,SP             ;and update DUTY LSB
39               lda   1,sp             ;get DUTY MSB
40               ldx   3,sp             ;get FRACDUTY
```

Code Listing 4.6: Variable Percent Duty Cycle Pulse Width Modulation using the TPM (*Continues*).

```
41                mul                ;multiply
42                stx   1,sp         ;store product integer part to DUTY MSB
43                add   2,sp         ;add fraction part to DUTY LSB
44                sta   2,sp         ;and update DUTY LSB
45                clra               ;for rounding if add produced a carry
46                adc   1,SP         ;add carry to DUTY MSB
47                sta   1,SP         ;update DUTY MSB
48                ldhx  1,SP         ;load DUTY for store and return value
49    ENDSETDUTY  sthx  TPMC0VH      ;write new duty cycle
50                lda   3,SP         ;return FRACDUTY in A
51                ais   #3           ;clean up stack
52                rts
```

Code Listing 4.7: (*Continued*) Variable Percent Duty Cycle Pulse Width Modulation using the TPM.

15.26 Hz at the selected bus clock rate and prescale value. Other frequencies are possible by changing the prescale value. Driver subroutine SETDUTY allows the duty cycle to be changed dynamically. The desired duty cycle is passed as a percentage in the carry flag C and index register X as a 9 bit unsigned fixed point value with 8 binary places (C.X). For example, C=0 and X=$80 specifies a 50% duty cycle and C=0 and X=$40 specifies a 25% duty cycle. C=1 specifies a 100% duty cycle regardless of the value in X (the duty cycle cannot exceed 100%).

Driver initialization subroutine INITTPM starts by writing the prescale value to TPMSC, with the CLKS field cleared to stop the TPM. The modulo value to be programmed is one less than the period; the defined period (decremented by 1 by an assembler expression) is loaded into HX and subsequently stored into the 16 bit TPM modulo register. Next, SETDUTY is called with C=0 and X=0, specifying an initial duty cycle of 0%. Timer channel 0 is enabled for positive polarity PWM with no interrupts by writing $24 to TPMC0SC. Finally, the timer is started by selecting the bus clock in TPMSC.

Timer subroutine SETDUTY can be called at any time to set the duty cycle to a desired value. The duty cycle is passed as a fixed point number representing the percentage of the period the signal is active (high, in this case). The basic function of the driver is to compute a value for the channel 0 value register in order to achieve the desired duty cycle. This value is given by TPMC0V=p_{Duty} * (TPMMOD+1), where TPMC0V is the 16 bit TPM channel 0 value, TPMMOD+1 is the TPM period, and p_{DUTY} is the percentage duty cycle passed as a parameter. When C=1, the duty cycle is set to 100% by writing the maximum value $FFFF to TPMCH0V. When C=0, the product of integer (TPMMOD+1) and fixed point value X must be computed, yielding a 16 point integer result. This product requires a 16-bit by 8-bit multiplication, which is accomplished by a series of 8-bit multiply and rounding operations. These operations are documented in the code.

4.6   CHAPTER PROBLEMS

1. List the three MC9S08QG4/8 modules that facilitate real-time processing. Describe a processing task for which each type is well suited.

2. Describe how each of the modules in the previous question could be used to blink a LED with a fixed period and 50% duty cycle.

3. What are the seven programmable periods for the RTI module using the internal clock reference.

4. Modify Code Listing 4-1 to allow for a real time period that is greater than a day.

5. What is the maximum delay that can be generated using the modulo timer?

6. Describe a repetitive task for which the non-blocking delay driver in Code-Listing 4.4 is appropriate.

7. In terms of code maintenance, why would the non-blocking delay driver in Code Listing 4.4 be advantageous even if a fixed processing delay is guaranteed in each period?

8. Describe a use for a pulse-width modulated signal generated with the PWM module.

9. Describe how you could use each of the three real-time modules (RTI, modulo-timer, and PWM) to blink an LED with a fixed frequency and 50% duty cycle? Which could be used if a variable duty cycle were desired?

Biography

DOUGLAS H. SUMMERVILLE

Douglas H. Summerville is an Associate Professor in the Department of Electrical and Computer Engineering at the State University of New York at Binghamton. He was a Assistant Professor in the Department of Electrical Engineering at the University of Hawaii at Manoa and a visiting faculty researcher at the Air Force Research Laboratory, Rome, NY and the Naval Air Warfare Center, Patuxent River, MD. He received the B.E. Degree in Electrical Engineering in 1991 from the Cooper Union for the Advancement of Science and Art, and the M.S. and the Ph.D. degrees in Electrical Engineering from the State University of New York at Binghamton in 1994 and 1997, respectively. He has authored over 35 journal and conference papers. He is a senior member of the IEEE and a member of the ASEE. He is the recipient of one service excellence and two teaching excellence awards, all from the State University of New York. His research and teaching interests include microcontroller systems design, digital systems design and computer and network security. Email: dsummer@binghamton.edu.